Toward the

of

And Two Additional Historical Studies

ROME'S FALL RECONSIDERED

HAY AND HISTORY

BY

Vladimir G. Simkhovitch

New York

THE MACMILLAN COMPANY

1925

PRINTED IN THE UNITED STATES OF AMERICA

Set up and printed. Published November, 1921.
Reprinted......December, 1925.

Press of
J. J. Little & Ives Company
New York, U. S. A.

PREFACE

PROBLEMS of history are problems of understanding. Three such problems of historical understanding are treated in this book.

The first study deals with the historical problem presented by the teachings of Jesus. The problem is —*why* such unprecedented teachings at *that particular time?* This study, therefore, deals, if you please, with the "fullness of time." To the scientific historian everything that has happened, happened in the "fullness of time" and to understand intimately and realistically that "fullness" is *the* task of history. In this particular study, therefore, we are endeavoring to understand the particular circumstances and conditions that make so great an historical event as the insight of Jesus historically intelligible to us.

In dealing with the gospel texts I have not attempted to utilize critical literature. Text-criticism to my way of thinking already presupposes a clear-cut understanding of the controlling factors in the historical situation.

"Rome's Fall Reconsidered" was first published in "The Political Science Quarterly" in 1916. After the essay was in print I was glad to hear that the famous German chemist, Justus Liebig, who contributed so much to agricultural chemistry, expressed in one of his agricultural works the idea that the decay of the Roman Empire was due to the decay of its agriculture. I certainly would have paid just tribute to Liebig had I been aware of it while writing my study. But I was not.

To the reader "Rome's Fall Reconsidered" may seem an obvious continuation of my "Hay and History" which was published first in "The Political Science Quarterly" in 1913. The general thesis is clearly developed there, yet it was another circumstance that led me to reconsider Rome's Fall.

In examining various data relating to the English Enclosures I became convinced that they were due not to the greater profit received by raising wool, but to the hopeless condition of English agriculture due to the exhaustion of the fields. This conviction is but briefly indicated in my Rome's Fall. The thesis was most convincingly established by a gifted student of mine, Miss Harriet Bradley.* In the meantime researches in the history of the productivity of human labor led me to read the Scriptores Rei Rusticæ, the ancient agricultural writers. Great, therefore, was my surprise to find Varro's plaintive descriptions of the abandonment of agriculture in Rome and of their turning fields into pastures. Varro's "contra leges et segetibus fecit prata" sounded remarkably like the familiar outcries of the early Enclosure movement. This similarity impelled me to reëxamine the decay of

* Harriet Bradley, *The Enclosures in England, Columbia University Studies in History, Economics and Public Law*, 1918.

the classical civilization. The results of this examination I present to the reader.

I am indebted to many intimate friends for their long-suffering patience in reading my studies in manuscript and for their valuable comments and criticisms.

VLADIMIR G. SIMKHOVITCH.

Columbia University,
October, 1921.

TABLE OF CONTENTS

Toward the Understanding of Jesus

TOWARD THE UNDERSTANDING OF JESUS

CHAPTER I

THE teachings of Christ are an historical event. Let us try to understand them historically. Without an historical understanding we have before us not teachings but texts. There is hardly a text in the four gospels that is not apparently conflicting with other texts. Yet an insight is won when the teachings of Jesus are viewed and understood historically.

The test of true understanding is to see in seeming contradictions but differing aspects of the same fundamental forces, to perceive in the endless expressions of life but one flow of life and to trace that flow to its sources. The test of true understanding is an understanding free from contradictions. So long as we find contradictions it is certain that what we hold in our hands are fragments; and though we may try to arrange them logically, the complete sphere of Jesus' own life and the life he preached we do not understand.

The gospels themselves contain practically nothing that throws light on Jesus' life as a whole. Little is to be found about his life and development before his ministry. Yet it is clear that when he entered upon his ministry he felt called to do so, and it is clear that such a mission develops slowly. What do we know of the long years while Jesus was thinking and feeling and praying, the years while the life was ripening which he afterwards preached and finally sacrificed? Under what circumstances he was developing, what he was doing, what influences impressed themselves upon his life and thought before he was thirty—what do we know about it? Nothing! The episode from Jesus' childhood, when he remained in the temple listening and asking questions of the learned men there, only emphasizes our lack of knowledge. For if Jesus in his childhood was so eager and mentally so keen, what was his mind doing during the eighteen or twenty years which followed that episode? Luke tells us "And Jesus increased in wisdom and stature, and in favor with God and man." [1] That is all we know about the growth and development of Jesus' life and mind. Was his inner life dormant or non-existent all these years? Did he not grow at all? Had his ideas no sources whatsoever, no development of any kind; were they utterly uncorrelated with the lives of his fellow men? What was Jesus, a phantom abstractly existing in a vacuum, or a historical personality really living and suffering in a given time and place?

There can be but one relevant answer to the question: Jesus was a historical personality. We all live and die and most of us are forgotten. Personalities

[1] Luke 2:52.

who are remembered, whom written records of human existence cannot overlook and our memory cannot forget, are personalities whose individual lives greatly affected many lives. A personality in other words acquires historical importance when it deals with the many, when its ideas, actions, words are understood by the many, affect the many. If a multitude gathers around one, it means that what the one is teaching is of interest to so many individuals that they form a multitude around him.

The more limited is our knowledge of the one, the more important is the light that may be shed by the many. The many seldom present difficult problems, for it is never very difficult to find out what in a given situation they had in common. What were their common conditions of existence, what were their common hopes, what were their fears, interests, purposes? Once we find that out, the reactions of the many are not difficult to understand. The particular historical conditions under which Jesus developed, lived, ministered and died are bound to help us understand his life and hence his teachings more intimately. How the Greeks or the Romans, the Gauls, the Goths or the Slavs at various times conceived and pictured to themselves Jesus and his teachings is an interesting problem in itself. It is the history of Christianity, it is the story of Jesus in the course of human history. The history of these interpretations of Jesus is a history of assimilations, in a sense a history of mankind. But it is not the history of mankind that interests us here. These interpretations can only confuse us. Nor are we interested in a composite picture of Jesus in history throughout the ages of faith. What we are searching for is that definite, concrete, historical Jesus who can give coherence to his teachings. Our quest is the historical truth. Let us therefore go to the documents; but let us be clear in our mind as to their value.

For historical truth is not a bundle of documents. Documents are the raw material, but not the structure. Historical truth is such a constructive insight into a given situation as to carry with it conviction of real life. Social life is then moving within its conditions of existence; and personalities, in their words and deeds, are correlated with their fellow men and appear in their historical, that is, their representative capacity.

CHAPTER II

In the year seventy after Christ the temple of Jerusalem was destroyed, Jerusalem was sacked, and the population either slain, crucified or sold into slavery. It is estimated that over a million and two hundred thousand perished. Josephus tells us about the destruction of Jerusalem that "the multitude of those that therein perished exceeded all the destructions that either men or God ever brought upon the world." [1]

The conventional history usually begins this war on August sixth of the year 66, when the Romans and other Gentiles were massacred by the Jews of Jerusalem. This date is so artificial that Mommsen for instance suggests A. D. 44 as the year from which the Jewish-Roman war might better be dated.

> It has been customary to put the outbreak of the war in the year 66; with equal and perhaps better warrant we might name for it the year 44. Since the death of Agrippa, warfare in Judea had never ceased, and alongside of the local feuds, which Jews fought with Jews, there went on constantly the war of Roman troops against the seceders in the mountains, the Zealots, as the Jews named them, or, according to Roman designation, the Robbers.[2]

[1] Josephus: Jewish Wars, VI, 9, 4.

[2] Mommsen: The Provinces of the Roman Empire, v. 2, p. 221-222. (New York, 1887.)

But to date the beginning of the revolt against Rome with the death of Agrippa in the year 44 is also quite arbitrary. For the revolt had been brewing and repeatedly breaking out here and there long before that. If we should follow the opinion of a contemporary historian, Josephus, we should have to date the beginnings back to the revolt of Judas, the Galilean, or Judas, the Gaulonite, to whose revolutionary activities and doctrines Josephus attributes all the ensuing misfortunes of the Jewish nation, which culminated in the destruction of Jerusalem and its temple. The occasion of that uprising was the census of Quirinius for taxation purposes in the year 6 A. D. Josephus tells us that one Judas, the Gaulonite, with a Pharisee named Saddouk, urged the Jews to revolt, both preaching that "this taxation was no better than an introduction of slavery, and exhorting the nation to assert its liberty." Josephus proceeds to inform us about these men and their doctrine:

> All sorts of misfortunes sprung from these men, and the nation was infected with this doctrine to an incredible degree; one violent war came upon us after another . . . . the sedition at last so increased that the very temple of God was burnt down by their enemies' fire.[1]

Toward the end of the same chapter he gives us some information about the so-called philosophy of Judas, the Gaulonite or the Galilean, as well as of his followers.

> These men agree in all other things with the Pharisaic notions; but they have an inviolable attachment to liberty, and say that

[1] Josephus: Antiquities, XVIII, 1, 1.

God is to be their only Lord and Master. They also do not mind dying any death, nor indeed do they heed the deaths of their relations and friends, nor could the fear of death make them call any man their master. And since this immutable resolution of theirs is well known to a great many, I shall speak no further about that matter; nor am I afraid that anything I have said of them should be disbelieved, rather do I fear that what I have said does not adequately express the determination that they show when they undergo pain.[1]

As a matter of fact the Jewish struggle for independence and the Zealot movement did not begin even with Judas the Gaulonite. Judas himself only continued the work of his father, Ezechias of Galilee,[2] who with his very large following was killed by young Herod when the latter was only the captain, στρατηγός, of Galilee under Hyrcanus, the ethnarch of Judea. That was in the year 46 B. C. Even then the Sanhedrin of Jerusalem must have had strong sympathies with Ezechias, for Herod was accused before that body for killing Ezechias and his followers, and he would have fared badly had not Sextus Cæsar, the Roman governor of Syria, requested from Hyrcanus Herod's acquittal.[3]

Nor does the rebellion of the Jews begin with Ezechias. The rebellion of the Jews against Rome rather begins with the power of Rome over the Jews; and in the same degree as the Roman power over the Jews increased, did the political reaction against that power, the revolution against Rome, increase and

[1] Josephus: Antiquities, XVIII, 1, 6.
[2] Schürer: Geschichte des Jüdischen Volkes im Zeitalter Jesu Christi, v. 1, p. 420. (4th ed., Leipzig, 1901).
[3] Josephus: Antiq., XIV, 9, 3-5. Jewish Wars, I, 10, 6-9.

spread. The Jewish revolutionists against Rome were called by the Romans bandits or robbers. Later they were called *sicarii,* "men with knives." The polite Josephus followed the Romans in calling them robbers; but whenever he tells us about the constant warfare, about either the Romans' or Herod's exploits against the robbers, it becomes clear that they are religious patriots who are fighting and dying for their country. So, for instance, Josephus describes one of Herod's expeditions against some Galilean robbers:

> Now these men slew the robbers and their families . . . and as Herod was desirous of saving some of them, he issued a proclamation to them . . . but not one of them came willingly to him, and those that were compelled to come preferred death to captivity. . . . And here a certain old man, the father of seven children . . . slew his children one after another. . . . Herod was near enough to see this sight and compassion moved him, and he stretched out his right hand to the old man and besought him to spare his children; yet did he not relent at all upon what he said, but reproached Herod on the lowliness of his descent, and slew his wife as well as his children; and when he had thrown their dead bodies down the precipice, he at last threw himself down after them.[1]

It is obvious here that we are dealing not with mercenary bandits, but with political and religious devotees who prefer death to submission. The Zealot movement, judging from Josephus's narrative, is of much older date than the revolt of Judas the Gaulonite, but that particular Galilean's uprising must have especially impressed itself upon the memory of men, for it is mentioned by way of illustration or characterization even in the Acts.

> After this man rose up Judas of Galilee in the days of the taxing [*i.e.,* the enrolment] and drew away much people after him: he

[1] Josephus: Jewish Wars, I, 16, 4.

also perished; and all, even as many as obeyed him, were dispersed.[1]

Still more important the outbreak becomes when we consider what happened at the same time. For it was this very Census of Quirinius and this very enrolment which, according to Luke, brought Joseph and Mary from Nazareth to Bethlehem where Mary gave birth to Jesus.[2] The chronology and details of Luke's narrative present many puzzles, important no doubt to the historian of dates and places, but not relevant in a history of ideas. The slight chronological discrepancies here we may overlook. For after all so far as influence and ideas are concerned, it does not matter whether the uprisings of Judas took place in the year 1 or in the year 7 after the birth of Christ. Certain it is that the great events under the shadow of which Jesus spent his childhood were memories of Herod's bloody rule, the annexation of Judea to the Roman province of Syria, and the revolt against Rome of Judas of Galilee.

The importance of Judas's uprising is attested to us by Josephus. The ideas for which Judas stood did not die with him, but were spreading and increasing till all of Judea and Galilee were in a veritable conflagration. Is it reasonable to suppose that Jesus paid no attention to what was going on around him? Is it reasonable to suppose that the souls of his fellow men, their ideas and ideals could be a matter of indifference to him?

You must remember that if there was a difference

[1] Acts 5:37.
[2] Luke 2:1-6.

between the Pharisees and the Zealots it was only in the method and the degree of their antagonism to Rome. The immediate followers of Judas grasped the sword as their answer to Roman taxation. But all the Jews in Jerusalem and throughout Judea resented the idea of paying tribute. Josephus tells us that they took the report of taxation "heinously," and that it took a great deal of persuasion on the part of the high priest Joazar to make them submit to the taxation.[1] It is clear, however, that the difference in attitude between the Zealots and the Pharisees was that the former resisted with the drawn sword, while the submission of the latter was but passive resistance, with a heart full of resentment but with an arm too feeble or a mind too cautious to grasp the sword. Hence the Pharisaic question, "Is it lawful to give tribute unto Cæsar, or not?"[2] The inquirer knew as well as Jesus how unpopular the answer Yes would be with a Jewish audience. Jesus answered, however, in the affirmative, pointing out that they have lost their independence, that on their tribute coin is the image of Cæsar. Hence there is nothing left but to "render therefore unto Cæsar the things that are Cæsar's."[3]

In the year 6 Judea was annexed to Syria; in the year 70 Jerusalem and its temple were destroyed. Between these two dates Jesus preached and was crucified on Golgotha. During all that time the life of the little nation was a terrific drama; its patriotic emotions were aroused to the highest pitch and then still more

[1] Josephus: Antiquities, XVIII, 1, 1.
[2] Matthew 22:17.
[3] *Ibid.*

inflamed by the identification of national politics with a national religion. Is it reasonable to assume that what was going on before Jesus' eyes was a closed book, that the agonizing problems of his people were a matter of indifference to him, that he had given them no consideration, that he was not taking a definite attitude towards the great and all-absorbing problem of the very people whom he taught?

In this setting, the Jewish nationalist could not separate religion from patriotism. Roman taxation, for instance, is certainly a purely political question, but Judas made a religious issue of it; and the Pharisaic interrogator of Jesus asked whether it was "lawful," that is, religiously permissible. Jesus therefore could not meditate about the religious problems of the people to whom he ministered without giving consideration to their engrossing political problem. That he had profoundly considered the problems of his day and wondered what the future contained for his people is shown by his reproach to the Pharisees:

> The Pharisees also with the Sadducees came, and tempting desired him that he would shew them a sign from heaven. But he answered and said unto them, When it is evening, ye say, It will be fair weather: for the sky is red. And in the morning, It will be foul weather to-day: for the sky is red and lowering. O, ye hypocrites, ye can discern the face of the sky; but can ye not discern the signs of the times? [1]

[1] Matthew 16:1-3.

CHAPTER III

CALLED upon to examine the origin and causes of the spreading resentment, of the fermenting revolution against Rome's rule, one curious circumstance is bound to attract our attention. This circumstance is that the Jews themselves petitioned Rome for Judea's annexation to the Roman province of Syria. Rome, on the other hand, did not grant the petition immediately. Only after years of Archelaus's misrule in Jerusalem was he finally deposed and Judea annexed in 6 A. D. Of course there were good reasons for the Jewish petition; the immediate concrete situation must have suggested precisely such action both on the part of Judea and on the part of Rome. But behind the immediate situation a vista is opened on the character and quality of Jewish political independence.

The events themselves are simple enough; judged by themselves they are insignificant; but valued psychologically, viewed as indications, what a light they throw upon Jewish nationalism and anationalism, upon Jewish political life with its dreams, its aspirations, its struggles and its fate. One has only to glance at the position of the heirs of Herod before Cæsar's throne; one has only to listen to the petitions and supplications of Herod's heirs and of the Judean ambassadors to realize that the political doom has long ceased to be a specter and a threat, but has been accepted by the

Jewish statesmen as an actual *status quo*, as a matter of fact, whether the plain people realize it in their every-day life or not.

Thus it seems that in their petitions they were haggling over minor terms and comforts; only details of submission appear to have worried them. In reality they were trying to save their culture and their religion. But why did the Jewish ambassadors demand provincial annexation? Why did eight thousand Jewish residents in Rome second Jerusalem's petition? Did they not prefer to be at least nominally independent? Fifty Jewish ambassadors were prostrated before the throne of Cæsar begging for annexation; the entire Jewish population of Rome was supporting these ambassadors and opposing the claims of Herod's heirs. Where then was Jewish patriotism, where the exclusive nationalism, clothed in all-consuming religious fervor? Fifty ambassadors were not likely to represent one particular clique; the entire Jewish populace in Rome could not be moved by considerations of sheer expediency. On questions of reason, feasibility, and expediency we divide; only on the most elemental emotions are we united. Hence their petition could not possibly have gone against those essentials which then constituted Jews as Jews; it could not go contrary to their religion and their nationalism.

Indeed it did not. Their supplications were dictated by austere and conservative religious nationalism. It was not for minor comforts they were bargaining. Rather did they feel that where the question at issue was between so-called political independence and religion, then indeed it was their religion, as they understood it, their Jewish culture that they could not possibly sacrifice. It was in reality a phase of the nationalistic struggle, although it took the curious form of a petition for annexation. If they should be managed by a Roman procurator, they hoped for complete cultural autonomy, and they expected to manage their own local affairs. Ruled by a Herodian prince, they were quite helpless to do so; for the Herodians, while nominally Jews, were striving hard to be culturally Romans. Naturally enough the cultural aspiration of their entire *entourage* was also Roman and anational; and this anationalism was insidious and widespread, especially in upper-class circles.

The Jews' petition for annexation was therefore to be an exchange of their sham political independence for very real cultural autonomy. In other words, complete independence looked to the more enlightened part of the population like a forlorn hope; and the struggle was waged for a home rule that would not infringe upon religious traditions. Statesmen they may perhaps have been, to follow these tactics; but they were certainly not philosophers. They did not realize that the growing religious and cultural conservatism and nationalism were an ideological expression of their political unrest; were but the spiritual flavor of their national and political struggle for independence. They did not realize that their religious culture and their political nationalism were so intimately tied up together that they could be served only by the sword. Hence it was most unlikely that cultural autonomy could really accept and adjust itself to the political downfall and the annexation as a province for which they were petitioning. Besides, so far as Rome was concerned, there was but one practical alternative. A Herodian government under Rome, offering no resistance to Rome, was precarious and undermining. It was tantamount to a complete cultural surrender.

Among the abuses of Herod, which the ambassadors quoted as reason for annexation, is the frank statement, that

> Herod did not abstain from making many innovations, according to his own inclinations. . . . That he never stopped adorning the cities that lay in their neighborhood, but that the cities belonging to his own government were ruined and utterly destroyed.[1]

Just how was Herod adorning the cities of the Gentiles? It is not uninteresting or unimportant. In Samaria Herod built

> a very large temple to Cæsar, and had laid round about it . . . the city Sebaste, from Sebastus or Augustus.

With similar temples to Cæsar he filled Judea, and when in honor of Cæsar

> he had filled his own country with temples, he poured out the like plentiful marks of his esteem into his province, and built many cities which he called *Cesareas*.[2]

In one of these Cesareas Herod also erected an

> amphitheater, and theater, and market-place, in a manner agreeable to that denomination; and appointed games every fifth year

[1] Josephus: Antiquities, XVIII, 2, 2.
[2] Josephus: Jewish Wars, I, 21, 2-4.

and called them . . . Cæsar's games, and he himself proposed the largest prizes upon the hundred and ninety-second Olympiad.[1]

Herod went farther still in emphasizing his adherence to so-called Greco-Roman culture. He built amphitheaters in Tripoli, Damascus and Ptolomeis, *agoras* at Berytus and Tyre, theaters in Sidon and Damascus.

And when Apollo's temple had been burned down, he rebuilt it at his own expense. . . . What need I speak of the presents he made to the Lycians and Samnians? or of his great liberality through all Ionia? . . . And are not the Athenians and Lacedemonians, the Nicopolitans and that Pergamus which is in Mysia, full of donations that Herod presented them with? [2]

As Herod's most splendid gift Josephus regards the endowment of the Olympic games, which were suffering much from lack of funds.

What favors he bestowed on the Eleans was a donation not only in common to all Greece, but to all the habitable earth as far as the glory of the Olympic games reached.[3]

As a matter of fact Herod even took part in these games himself.

These activities of Herod are obviously too strenuous, too consistent to be casual. Inwardly anything but a Roman gentleman, he took the world-culture, Hellenism, for his ideal, and made outward assimilation to that culture his ardent endeavor. He was far from being a unique specimen in Judea. Many felt as he did, but they belonged to the upper classes and were certainly a small minority. The bulk of the popu-

[1] Josephus: Jewish Wars, I, 21, 8.
[2] Josephus: Jewish Wars, I, 21, 11.
[3] Josephus: Jewish Wars, I, 21, 12.

lation resented and resisted the Greco-Roman culture; they resisted it religiously as sacrilege and nationalistically as treason.

It was a tangible incident of precisely such nature which led to the break, to a revolt and a petition for annexation.

Josephus reports the speech which Herod made to the people of Jerusalem when he was about to rebuild their temple. He told them what made his undertaking possible:

> I have had peace a long time, and have gained great riches and large revenues, and, *what is the principal thing of all,* I am at amity with and well regarded by the Romans, *who, if I may say so, are the rulers of the whole world.*[1]

Subordination to Rome, however, was emphasized in more than speeches. The very temple of Jerusalem was to bear witness thereto. A large Roman eagle made out of gold at vast expense was erected over the principal gate of the temple. Since any kind of image was forbidden to Jews by the law and the prophets, that Roman eagle was not exactly cherished. Resistance against Herod just then was useless. When Herod's health began to fail, however, the Jews started an agitation to remove from the temple the eagle which in their eyes was both a sacrilege and a national insult. The leaders of the movement were the most eloquent two Jews of their time: Judas, the son of Saripheus, and Matthias, the son of Margalothus, both teachers of the law. They realized that Herod would punish their deed with death. But they

Josephus: Antiquities, XV, 2, 1.

felt that those who die for such a deed "will die for the preservation and observation of the law of their fathers and will also acquire everlasting fame and commendation." [1] The eagle was pulled down and cut to pieces by a number of young men under the leadership of this Matthias and Judas, and Herod ordered all of them to be burned alive.[2]

These men were honored by the Jews as martyrs. When in the course of time Herod died and Archelaus succeeded him, Archelaus, pending Rome's confirmation of his succession, was very anxious to please the people and avoid annexation. The people demanded lower taxes, lower duties on commodities, freedom for prisoners. All these demands Archelaus gladly granted. But then the people began to mourn the rebels whom Herod had burned. Let us quote Josephus again:

> They lamented those that were put to death by Herod, because they had cut down the golden eagle that had been over the gate of the temple. Nor was this mourning of a private nature, but the lamentations were very great, the mourning solemn, and the weeping such as was loudly heard over all the city, as being for those men who had perished for the laws of their country and for the temple. They cried out that a punishment ought to be inflicted for these men upon those that were honored by Herod.[3]

Not being able to appease the multitudes, Archelaus resorted to force. About three thousand Jews were slaughtered by his soldiers. It was this incident which led to a general uprising and an intervention of the

[1] Josephus: Antiquities, XVII, 6, 2.
[2] Josephus: Antiquities, XVII, 6, 3-4.
[3] Josephus: Jewish Wars, II, 1, 2.

Roman forces, and to the deputation from Jerusalem which petitioned annexation.

Should an impression be gained that the Herodians were responsible for Hellenizing the people and Romanizing the commonwealth, this impression is completely out of focus and erroneous. The Herodians themselves were pawns in the game, mere incidents that may serve as illustrations. The Hellenistic tendency, the tendency toward world culture and toward a Judaic anationalism filled the pages of Jewish history, not only long before the Herodian dynasty, but even long before the Hasmonean ascendency. In fact it was the popular and religious reaction to that very tendency that culminated in the Maccabean struggles.

The prelude to the Maccabean struggles introduces us to an educated upper class, Hellenized and Hellenizing, and to an opposing party called "the pious" or "the Chassidim." It was not the Chassidim who had the upper hand. The government was in the hands of the Hellenistic party. The high priest was a certain Jason, who was hardly behind Herod in his "cultural" tendencies. He, too, sent many gifts to pagan festivals, such as the sacrificial festival of Hercules at the games in Tyre. He, too, erected a Greek gymnasium under the castle of Jerusalem; and the author of the Second Maccabees reports to us that he caused the noblest of the young men to dress like Greeks.

> And thus there was an extreme of Greek fashions, and an advance of alien religion, by reason of the exceeding profaneness of Jason that ungodly man and no high priest: so that the priests

had no more any zeal for the services of the altar, but despising the sanctuary, and neglecting the sacrifices, they hastened to enjoy that which was unlawfully provided in the palaestra after the summons of the discus; making of no account the honors of their fathers, and thinking the glories of the Greeks the best of all.[1]

Hellenism was rapidly encroaching upon Judaism, and the Hellenistic party had full sway in Jerusalem. Frankly they were none too proud of being Jews. They conspired openly with Antiochus Epiphanes against those who held fast to the traditions, and encouraged him to accelerate the Hellenization and abbreviate its process. He prohibited the exercise of Jewish rites, on pain of torture and death; he forbade the Jews to remain Jews, to worship the God of their fathers, and by force compelled them to sacrifice to the gods of Olympus and of the Hellenic world.

When thus a war of extermination was waged by the Syrian king against the Jews; when no other alternative was left them but to sacrifice either their lives or their religion, then they arose determined to defend both in an unequal struggle. Under the leadership of Matthias and, after his death, of his son Judas, the Maccabee, the Jews inflicted severe punishment upon the generals of Antiochus. Wherever the victorious arms of the Maccabeans went, they swept before them all Hellenism and anationalism. The Maccabean family established themselves first as popular leaders and later as a theocratic dynasty of high priests and rulers of the people, the Hasmonean dynasty.

Such is the epitome of a phase of the struggle which lasted decades. Most of the details, of course, are

[1] II Maccabees 4:13-15.

of no interest to us. And yet there are pages in the history of these struggles which are important for later reference and which we should remember. First of all, Judas the Maccabee, whilst struggling against heathendom, is forced to seek an "alliance" with Rome.[1] For it became early enough quite clear that no amount of courage could avail the little nation in the long run against the superior strength of Syria. So we find Simon, the brother of Judas, who succeeded him in leadership, again sending ambassadors to Rome in 139 B. C., who brought with them rich gifts and sought the renewal of Judas' covenant of friendship. Thus already in the days of their struggle for independence from Syria the Jews were obliged to seek the protection of Rome. This same protection led not very much later to intervention and dependence.

Another detail that should be borne in mind is that Judas Maccabaeus, even in the days when fortune smiled upon him and victory accompanied his arms everywhere, could not undertake to secure Judaism in either Galilee or Gilead. There the Gentiles were so numerous and so strong that the early Maccabees did not even undertake to Judaize these provinces. The first book of Maccabees and Josephus[2] inform us that Judas went to Gilead with one army and sent his brother Simon with another army three thousand men strong into Galilee. After many battles against the heathen in Galilee and as many victories, Simon gathered all the Jews in Galilee with their households

[1] I Maccabees 8; Josephus: Antiquities, XII, 10, 6.
[2] I Maccabees 4:60-61. Josephus: Antiquities, XII, 7, 7.

and their goods and convoyed them amid great rejoicing to Judea where they could be secure.

One hundred and fifty years later, when Jesus lived among the people of Galilee, it was of course a different Galilee. Judaism was strong there and at times peculiarly intolerant of Roman domination, as the rebellion of Judas, the Gaulonite, or even that of his father proved. But there is also little doubt about the large Gentile population in Galilee, much larger than in any part of Judea proper. Where two races are living side by side with differing traditions and differing religions, two social phenomena can as a rule be observed: greater mutual understanding than elsewhere and greater tolerance under ordinary circumstances, for the strangers are not strange to them; but in times of excitement, greater antagonism, race hatred and general intolerance, for strangers are near at hand. Thus in Galilee, where Jews and Gentiles came in close contact, there was the basis for relations more antagonistic as well as more friendly. When the Jew was friendly he was likely to speculate and wonder whether after all his Heavenly Father were not the father of the Gentile fisherman and farmer as well. When, however, the Jew of Galilee was unfriendly, the very proximity and daily contact with the Gentile must have made him peculiarly jealous of Jewish independence. For Jewish independence meant Jewish ascendency in a mixed population, while Jewish dependence involved not only national degradation, but also particular and immediate personal degradation in the Jew's relative position to his Gentile neighbor.

Among the upper classes of Jerusalem, Philhellenism was probably never completely stamped out. It is well to remember that assimilation and admiration for foreign habits and ideas are much more likely to be found on the top than at the bottom of society; for what characterizes the lower and humbler strata is their traditionalism. Here is another detail of the Maccabean struggle that may serve as an illustration; for the very chronicles of Maccabees emphasize indirectly the fact that it was difficult to eradicate Hellenism. The first book tells us about Jonathan, the brother of Judas:

> The sword was now at rest in Israel, and Jonathan dwelt in Michmash; and he began to judge the people, and drove out the ungodly from Israel.[1]

The ungodly were of course the Jews with Hellenistic tendencies. Yet how difficult it was for him to drive out Philhellenes is shown by the fact that Jonathan had to live in Michmash. He lived in Michmash because Jerusalem was at the time in the hands of that very ungodly Hellenistic party.

True, soon enough Jerusalem was in the hands of the Maccabeans, but no sooner did the Hasmonean dynasty completely establish itself than it, too, began to follow the trend toward the world culture. Thus we find John Hyrcanus abandoning the Pharisees, the strictly orthodox party, and associating with the Sadducees. Neither the Pharisees nor the Sadducees were a sect with static dogmas, as text-books of theology are likely to present them to us. Rather do both sects represent potential tendencies and viewpoints. The

[1] I Maccabees, 9:73.

Pharisees accepted traditional interpretations of the law, for they were traditionalists. The Sadducees accepted the national and religious minimum; the law, but not the added traditions. They were the rich upper class; the populace were with the Pharisees.

> The Sadducees are able to persuade none but the rich, and have not the populace obsequious to them; but the Pharisees have the multitude on their side.[1]

The law itself, obviously could never be so culturally isolating as the law plus the entire body of sanctified traditions. The well-to-do liberals could easily be persuaded to drop the traditional additions and interpretations, but not so the populace. Neither was the populace in the habit of giving themselves Greek names. But all the sons of Hyrcanus have Greek names: Aristobulus, Antigonus, Alexander. To be sure, as a high priest Aristobulus had use for a Hebrew name as well, which happened to be Yehuda—Judas. This king, according to Josephus, either so favored Greek ideas that he was known as a lover of Hellenism, as a Philhellen, or actually adopted the title Philhellen.[2] Aristobulus's successor and brother, Alexander Jannaeus, even introduced bilingual coins with the two inscriptions in Greek and Hebrew, and incidentally adopted the title Melek—βασιλεύς.

All this tends to show that the Hellenistic tendency was not of Herodian making. It existed fully as strongly in the pre-Maccabean period; it was checked by the nationalistic and religious revolt of the Macca-

[1] Josephus: Antiquities, XIII, 19, 6.
[2] Antiquities, XIII, [illegible] 3.

beans; it revived again under the Hasmonean dynasty. And as the little Jewish kingdom was becoming more and more a dependency of Rome, two tendencies were rapidly developing; that of submission to Rome and cultural assimilation among the upper class, and that of growing nationalism and religious orthodoxy. The nationalism and religious orthodoxy became one and indivisible, yet the accent was on the religion, for tradition was bound up with religion. Tradition was religious, and what else after all was nationality but the sum total of traditions?

Romanization threatened the very life of their tradition; it interfered with their religion. A Herodian prince ruling by the grace of Rome was sure to interfere much more than a Roman administrator. At least they thought so. They wanted independence; but if no independence was to be had, the next best thing was cultural home rule under a Sanhedrin of their own choosing, autonomy that would guarantee them their own religious traditions. Such autonomy was unthinkable under a Herodian prince. It was quite conceivable under a Roman governor. Hence their petition for annexation to Syria. Interesting it is that the vicissitudes of their national history, a long, long history that was dating back to their Babylonian captivity, taught the Jews to consider themselves primarily a religious entity; interesting it is that the Jews themselves were petitioning to be permitted to render to Cæsar what is Cæsar's for the sake of being free to give to their God what is God's.

CHAPTER IV

THE annexation of Judea to the province of Syria, in spite of the possibility which it offered of greater cultural autonomy, could neither solve the problem nor save the situation. Granting even that orthodoxy in Jerusalem had a freer hand under a Roman procurator than it could have had under a Herodian prince, that more tenacious orthodoxy was in itself but a reaction against the encroaching national doom. By annexation the national doom was being not averted but consummated. True enough, any Roman Pontius Pilate would have let the Jews have their own way in religious matters. He would have washed his hands of them, while a Herodian king would wash his hands in the blood of his Jewish adversaries. But what was nationalistic orthodoxy gaining? Subjectively and psychologically the Jews were losing, more irrevocably than ever.

Where cultural assimilation preceded political and territorial absorption by Rome, the final act was felt but little. The death of a nation was made easy. The process of assimilation involved in fact a cultural compromise. The Romans themselves were culturally proselytized by Greece, by Egypt, by Mithraism, even by Judaism. Assimilation involved to some extent an exchange of cultural concepts. Jewish proselytism was but an incident of assimilation, not of Jewish nationalism. The Roman lady who became converted to Judaism and sent money to the temple remained a Roman lady. She might have chosen to worship Isis without becoming an Egyptian; she chose Jehovah without ceasing to be a Roman. Proselytism without national absorption was already a first step to assimilation, for it involved the denationalization of a national religion.

The process of Jewish assimilation was cut short in Judea by dramatic political events and the nationalistic reaction of the masses. The brutal aggression of Antiochus Epiphanes put an end to all assimilation and caused the Maccabean revolt. True enough, the cultivated and educated Jews realized quite well that they dealt with Rome, the ruler of the habitable earth. Whether a Herodian prince, or a Josephus, or a high priest like Joazar—any one of them knew what the Roman Empire was, knew that a conflict with that Empire could end in but one way. But the plain people knew only their traditional religion and glimpsed but vaguely the insuperable power of Rome.

Now that Rome was establishing herself firmly and frankly as Judea's avowed lord, the increased national feeling, the bitter national antagonism of the Jews was equally frank. The religion of their forefathers became the unfurled banner of a nation at bay. From now on, whether in passive resistance or in open rebellion, the only lord and master they recognized was the Lord of Hosts, the God of Abraham, Isaac and

Jacob, with whom they were in covenant, and who must send the great Deliverer to save his people in their hour of need.

Greater and greater became the pressure; greater and greater grew that need. Where was the Messiah? Would he come in the future? Oh, but there was no longer any future; it was then and there that he must come. Yea, to save his people he must have come already, must be among them, only unrecognized, unknown to them—Messiah, the anointed of God, the Christ.

Shall we now ask the question under what influences Jesus developed; what problems absorbed him before he began his ministry? Or is such a question superfluous? The central problem of his people was so enveloping that we can take for granted that Jesus' religious and intellectual life revolved around it, and that his own development consisted in the gradual solution of this very problem. To repeat, at the given time there was but one problem for the Jews—a single, all-absorbing national problem, that became under the circumstances *the* religious problem as well. It was the problem of existence, the problem of escape from certain annihilation.

One was the problem, but the solutions were several. Clearly the Jewish nationalists and the Herodians could not possibly agree upon the same solution. Even the religious nationalists of the time differed considerably. Yet in spite of all their differences as to method, their hope was the same. This hope was the national salvation, and their reliance was upon Messiah, the

Christ, the anointed King. Do you remember the song of Zacharias?

Blessed be the Lord God of Israel; for he hath visited and redeemed his people, And hath raised up a horn of salvation for us in the house of his servant David; As he spake by the mouth of his holy prophets, which have been since the world began: That we should be saved from our enemies, and from the hand of all that hate us; To perform the mercy promised to our fathers, and to remember his holy covenant.[1]

Faith in the immediate national Deliverer was the great need. That faith, though it attached itself to God's own promises, indicated in the law and the prophets, really opened a flood-gate of new religious interpretations, and new religious beliefs. The law and the prophets were given and standardized; they contained no detailed information about the great need, the actual means of deliverance and salvation. Here was a free region for mystic, religious and political speculation. Such speculation could not be standardized. It was like a set of popular supplements to the existing religion. Being of immediate significance, offering solutions of the immediate great problem, the free supplements had great weight. Of course the law was observed and revered traditionally, but interest was centered on these popular additions to the canonic scriptures. In scrutinizing the future, they were reinterpreting the past. Attached to the traditional religion, they were yet inevitably modifying the very law. Not only were Messiah and his kingdom an immediate political necessity and therefore the center

[1] Luke 1:68-74.

of interest, but theologically they fulfilled God's own part in the covenant with his chosen people. The old law laid out all the paths of conduct for its people. It never undertook, however, to regulate the ways of God. Interest was now concentrated upon these very ways of God, a realm offering unlimited freedom to new vision, new insight, new interpretation.

Intellectual, political and spiritual life was heightened and intensified. That is characteristic of all critical periods. Do not let us assume that orthodoxy was weakened. Not at all. There was now room for deeply religious heterodoxy, but orthodoxy itself became much intensified. The tension bordered on hysteria; as is indicated in the eschatological literature of the time and by the prevalence of nervous maladies among the people in the days of Christ's ministry.

This great nervous strain was part of the crisis. It is precisely such a crisis that leads the many to the border of hysteria or to nervous anomalies of one kind or another, and that leads the few to the most extraordinary social, intellectual and moral achievements. There should be nothing mystical about the trite observation that every crisis produces its great men. The fact is that under ordinary conditions of existence, when we are quite sane and safe, we are using but a small fraction of our potential intellectual and emotional powers. It is precisely such social strain produced by a crisis that increases not our potential capacity, but the percentage of capacity at which we are actually working, thinking, feeling. Such a crisis, while greatly increasing numerically the broad base of the intellectually and emotionally active members of society, quickens as well the activities of the individual, and further heightens the individual lives through their manifold interreactions. Greater achievement in both quantity and quality is almost inevitable. All dimensions are enlarged. Creative ability is enlarged; destructive folly is enlarged; all human activities, all elements of friction are increased for good and for evil; and the scale must be larger for the outstanding personalities who are to marshal the enlarged forces of life.

All dimensions being enlarged, single figures are not outstanding unless they are of heroic size. Hence they tower long afterwards over life's subsided flow, when humanity is again resting in routine existence from its mental strain or physical exhaustion. Conditions that call for intensified life with its ecstasy and hysteria, and its greater mental effort, are in their very nature inimical to all routine orthodoxy, political, religious or social. For orthodoxy is in its essence an established routine, and a crisis means exactly that the routine is endangered. Orthodoxy is the standardized organization, the delimitation of the flow of life at its low average level; it cannot hold within its banks the rushing freshets of a quickened life. Who in these troubled waters will undertake to discover where orthodoxy ends and heterodoxy begins?

The entire literature of the time is a fragmentary expression of this quickened life of the nation. The records of every Messianic hope contain a preamble somewhat similar to the especially well phrased passage in the Second Esdras, which, although written after Christ, expresses concisely the spirit of the constant Jewish question.

> All this have I spoken before thee, O Lord, because thou hast said that for our sakes thou madest this world. As for the other nations, which also came from Adam, thou hast said that they are nothing, and are like unto spittle: and thou hast likened the abundance of them unto a drop that falleth from a vessel. And now, O Lord, behold, these nations, which are reputed as nothing, be lords over us, and devour us. But we thy people, whom thou hast called thy firstborn, thy only begotten, and thy fervent lover, are given into their hands. If the world now be made for our sakes, why do we not possess for an inheritance our world? How long shall it endure? [1]

This in the main is the preface to the entire vast popular literature, political and prophetic, which covers a period of about three centuries, and of which but sample specimens survive. Conceived at different times under varying influences and conditions the character of the Messiah varied. A century or so before Christ Messianic quality was attributed to the early Maccabean leaders; a century after Christ the last great rebel leader Bar-Kochbah was viewed as the Messiah. Conceived under different oppressions, contemplated from different viewpoints, the scope and character of the Messianic kingdom differed widely. It will not be in keeping with our purpose to go through the entire gamut of tones and variations of salvation which this literature, in so far as it survives, offers us. It suffices that all this literature has one common pur-

[1] II Esdras 6, 55-59. The Apocrypha, Revised Version, 1894. See IV Esdras in Charles: Apocrypha and Pseudoepigrapha of the Old Testament, v. 2, p. 579.

pose: finally, somehow, it saves the Jew; the promises that it holds out to the Gentile world are less encouraging. Let us quote, for example, from the Assumption of Moses, written in all probability about A. D. 7-29. Like most of the Messianic literature it is replete with all kinds of heavenly signs, such supernatural signs as were demanded and made the criterion of Jesus's Christhood.

And the earth shall tremble; to its confines shall it be shaken: and the high mountains shall be made low and the hills shall be shaken and fall. And the horns of the sun shall be broken and he shall be turned into darkness; and the moon shall not give her light and the circle of the stars shall be disturbed. And the sea shall retire in the abyss, and the fountains of water shall fail and the rivers shall dry up. For the Most High will arise, the Eternal God alone, and he will appear to punish the Gentiles, and he will destroy all their idols. Then thou, O Israel, shalt be happy and thou shalt mount upon the neck and wings of the eagle,[1] and they shall be ended and God will exalt thee. . . . And thou shalt look from on high and see thy enemies in Gehenna and thou shalt recognize them and rejoice. And thou shalt give thanks and confess thy Creator.[2]

In the so-called Psalter of Solomon, written after Pompey's invasion of Judea somewhere between 63 and 48 B. C., the tenor is more or less the same, except that the hope is centered on a king of the House of David.

Behold, O Lord, and raise up unto them their king, the son of David, at the time in which Thou seest, O God, that he may

[1] *I.e.*, triumph over Rome.
[2] The Assumption of Moses, 10, 4-10. R. H. Charles: The Apocrypha and Pseudoepigrapha of the Old Testament. Oxford, 1913. V. 2, p. 420-421.

reign over Israel, Thy servant. And gird him with strength that he may shatter unrighteous rulers, and that he may purge Jerusalem from nations that trample her down to destruction.[1]

As in practically all the literature of this type, the Messianic hope and prayer spring from desperate political conditions no longer to be borne.

The lawless one laid waste our land so that none inhabited it. They destroyed young and old and their children together. In the heat of his anger he sent them away even unto the west. And (he exposed) the rulers of the land unsparingly to derision.[2]

The King of the house of David (as it was written under the non-Davidic Hasmonean dynasty) will crush the Gentiles, chastise the sinners, cleanse Israel and send the Messiah, the *χριστὸς κύριος*, the Lord's anointed, the king anointed of the Lord.[3] Here, as in all pre-Christian documents, the chief function of the Messiah is the overthrow of the oppressors, the crushing of the ungodly powers.

But there are also other notes in this literature of woe and hope. While in the main the problem of the time was to rid the nation of foreign oppression, the very familiarity with the omnipresent Gentiles was tending to undermine racial exclusiveness. Literary expression is naturally more conservative than actual life. For faltering and hesitating is the surrender of the literary tradition to life. It surrenders indirectly and incompletely. Sacred to tradition were the curses heaped upon the heads of the Gentiles; profane, secu-

[1] The Assumption of Moses, 10, 4-10. R. H. Charles: The Apocrypha and Pseudoepigrapha of the Old Testament. Oxford, 1913. V. 2, p. 649.

[2] Charles: Apocrypha. Oxford, 1913. V. 2, p. 648.

[3] *Ibid.*, p. 650-651. Psalms 18:8, 17:36.

lar and new were the growing familiarity, the enforced intimacy with the Gentile world. This new familiarity could not but affect the traditional and narrow outlook. In some of the popular literature, as, for instance, in "The Testaments of the Twelve Patriarchs," salvation is promised not only to the Jews, but also to the Gentiles, who are to be saved through Israel. On the other hand, the conduct of a good Gentile will be the standard by which Israel will be judged by the Lord. "And he shall convict Israel through the chosen Gentiles, as he convicted Esau through the Medianites who loved their brethren." [1]

The apocryphal literature went even farther; in its happier moments it cherished the visions and prophecies of Isaiah of eternal and universal peace and the future brotherhood of mankind. Incidentally how many of the beautiful sentiments in Virgil, in his Georgics and his Eclogues, especially in the fourth, so-called Messianic Eclogue, are copied outright from the Jewish Sybillines, who interpreted to the Greco-Roman world the old visions of Isaiah!

To some Jews of the time the visions of Isaiah were more than prophetic memories and quotations. There on the brink of war, they were benedictions of peace; on the threshold of death they were songs of love and life. Songs of love and life when most hearts were filled with mortal fear. How could fear-oppressed hearts listen to such songs? They could not. Only by hatred greater than their fear could that mortal fear be overcome. Hatred overcame their fear of

[1] Testament of Benjamin, 9:2, 9:10. Charles: Apocrypha. Oxford, 1913. V. 2, p. 358-359.

death, and hastened them into the arms of death. The struggle with Rome meant death. Was there no other way, no other solution?

Desperate was the external situation, desperate the inner pain of souls searching for a way out, instinctively reaching towards light and life. Thus all hope and aspiration were centered in the coming of a Christ whose mission is so wonderfully expressed in Luke: "To give light to them that sit in darkness, and in the shadow of death, to guide our feet into the way of peace." [1]

[1] Luke 1:79.

CHAPTER V

In the year 70 the national tragedy was consummated. The temple was burned and Jerusalem destroyed; its inhabitants were delivered unto the sword, crucified, sold into slavery and scattered to the four corners of the earth. So long protracted was the tragedy that Jesus' whole life and ministry occurred in the midst of it. The events of life do not come to us named and labeled; neither did Judea's life on the eve of its great historical catastrophe carry banners spelling "tragedy." But even a superficial glance at Jesus' life shows us the imminence of the disaster, and how concretely Jesus' life was bound up with the political destiny of Judea. For was not Jesus born in the days of the tax-enrolment? Did not in all probability the same tax-enrolment start the rebellion of Judas the Gaulonite? Did the battle-cry of Judas, "No tribute to the Romans," ever die out in Jesus' lifetime?

Multitudes followed Jesus. Shall we assume that his message was in no wise related to the paramount interest of the people? What did Jesus mean when he reiterated that he was sent to save the lost sheep of Israel? What did his followers have in mind when they perceived in him their Savior, their Messiah, their Christ? What was Messiah's function, what did the people of the time expect from their Messiah?

They expected their national salvation. What that national salvation meant was clear enough. Luke states it: "That we should be saved from our enemies, and from the hand of all that hate us." [1] He repeats it a few verses later: "That he would grant unto us that we, being delivered out of the hand of our enemies, might serve him without fear." [2]

Now when one talks about national enemies, one is talking about a given historical moment. It was therefore about a given and dreaded historical moment that Christ was speaking when he said:

> O Jerusalem, Jerusalem, thou that killest the prophets and stonest them that are sent unto thee, how often would I have gathered thy children together, even as a hen gathereth her chickens under her wings, and ye would not.[3]

And we see clearly a definite historical moment when we read:

> See ye not all these things? verily I say unto you, There shall not be left here one stone upon another, that shall not be thrown down.[4]

The inevitable end of the tragedy towards which the children of Israel were so swiftly tending was only too obvious. "And when ye shall see Jerusalem compassed with armies, then know that the desolation thereof is nigh." [5] Or as Mark states it:

> But when ye see the abomination of desolation . . . standing where it ought not (let him that readeth understand), then let them that be in Judea flee to the mountains: And let him that

[1] Luke 1:71.
[2] Luke 1:74.
[3] Matthew 23:37.
[4] Matthew 24:2.
[5] Luke 21:20.

is on the housetop not go down into the house, neither enter therein, to take anything out of his house: And let him that is in the field not turn back again for to take up his garment. But woe unto them that are with child, and to them that give suck in those days! And pray ye that your flight be not in the winter.[1]

I have always found that it takes an enormous amount of learning to get away from the most obvious and simple truth. So our modern theologians are explaining this statement eschatologically; that is, they see in it a prophecy of the end of the world. If it refers to the end of the world what difference does it make whether that end is to come in the winter or in the summer? Such obvious misinterpretation of this text indicates a complete lack of understanding of other texts. For indeed no understanding of the sayings of Christ is at all possible without at least a rudimentary insight into the historical background.

If we do not have before us the clear perspective of events which are inevitably coming unless the nation change its mind, how can we understand the following passage:

There were present at that season some that told him of the Galileans, whose blood Pilate had mingled with their sacrifices. And Jesus answering said unto them, Suppose ye that these Galileans were sinners above all the Galileans, because they suffered such things? I tell you, Nay; but, except ye repent, ye shall all likewise perish.[2]

Generally speaking, repentance, the Greek μετάνοια, a change of mind, has to be and can only be individual,

[1] Mark 13:14-18.
[2] Luke 13:1-3.

personal. Yet it is not with an individual, but with a national situation that Jesus was here clearly dealing. The Galilean patriots whom Pilate had slain were not sinners above all sinners; they were average representatives of the nation as a whole. They were a good sample and so was their fate. They perished, and the entire nation will perish if its mind is not changed.

It is truc that Christ's clear insight was not shared by his contemporaries. The populace could not see where their Pharisee and Zealot leaders were leading them and what fate they were preparing for themselves. The greater was the sorrow of Jesus; for perdition was in full sight yet hidden from their eyes.

> And when he was come near, he beheld the city, and wept over it, Saying, If thou hadst known, even thou, at least in this thy day, the things which belong unto thy peace! but now they are hid from thine eyes. For the days shall come upon thee, that thine enemies shall cast a trench about thee, and compass thee round, and keep thee in on every side, and they shall lay thee even with the ground, and thy children within thee; and they shall not leave in thee one stone upon another; because thou knowest not the time of thy visitation.[1]

Such texts not only invite examination of the concrete historical background; they actually supply, though in a fragmentary way, the very incidents of the historical situation. Does not the fourth gospel give us *in nuce* a complete insight into the entire situation by telling us what Caiaphas and the chief priests thought? "If we let him thus alone, all men will believe on him: and the Romans shall come and take away both our place and nation." [2] So they decided that it is expedi-

[1] Luke 19:41-44.
[2] John 11:48.

ent "that one man should die for the people, and that the whole nation perish not." [1]

The primary concern of the Pharisees and priests was also the fate of the nation. The Pharisees could probably have overlooked the heresies in Christ's religious teachings, as they overlooked those of the Sadducees, who denied such traditional canons as the immortality of the soul. The great and fundamental cleavage was constituted by Christ's non-resistance to Rome. Of course they could not use that as an accusation when they were seeking his condemnation at the hands of a Roman procurator, and they had to invent some other charges.

Even the Roman procurator seems to have had an insight into the situation, for he exerted himself to save Christ. According to the account in Luke, Herod, too, whose rule was not of the gentlest, found also no fault with Christ. For while neither of them of course understood Christ, they did understand that he was against rebellion.

> And Pilate, when he had called together the chief priests and the rulers and the people, said unto them, Ye have brought this man unto me, as one that perverteth the people: and behold, I, having examined him before you, have found no fault in this man touching those things whereof ye accuse him: No, nor yet Herod: for I sent you to him; and lo, nothing worthy of death is done unto him.[2]

But Pilate could not persuade them, and because of the tumult[3] he did not dare resist them. In so tense

[1] John 11:50.
[2] Luke 23:13-15.
[3] Matthew 27:24.

a situation he feared to provoke an outbreak of the rebellion.

Pilate offered to release Jesus because of the Passover feast. It was the exercise of a customary prerogative.

> And they cried out all at once, saying, Away with this man, and release unto us Barabbas: (Who for a certain sedition made in the city, and for murder, was cast into prison.) Pilate therefore, willing to release Jesus, spake again to them. But they cried, saying, Crucify him, crucify him.[1]

Mark's information about Barabbas is perhaps more specific.

> And there was one named Barabbas, which lay bound with them that had made insurrection with him, who had committed murder in the insurrection.[2]

Thus Christ was delivered unto his enemies and the rebel leader Barabbas was released. The patriots had won the day. They knew not what they were doing, nor realized that they were sealing the fate of their nation. To Jesus, however, it was quite clear; hence when the women of Jerusalem followed him on the way to Golgotha bewailing and lamenting him he turned to them and said:

> Daughters of Jerusalem, weep not for me, but weep for yourselves, and for your children. For, behold, the days are coming, in the which they shall say, Blessed are the barren, and the wombs that never bare, and the paps which never gave suck. Then shall they begin to say to the mountains, Fall on us; and to the hills, Cover us.[3]

[1] Luke 23:18-21.
[2] Mark 15:7.
[3] Luke 23:28-31.

CHAPTER VI

THE vision of the inevitable consequences of the brewing rebellion—was it Christ's unique insight, shared by no one? Hardly so. Many intellectuals probably foresaw and feared the outcome, but they felt powerless against the national passion. Interesting as an illustration is what Josephus tells us about himself.

> I therefore endeavored to put a stop to these tumultuous persons, and persuaded them to change their minds; and laid before their eyes against whom it was that they were going to fight, and told them that they were inferior to the Romans not only in martial skill, but also in good fortune; and desired them not rashly, and after the most foolish manner, to bring on the dangers of the most terrible mischief upon their country, upon their families, and upon themselves. And this I said with vehement exhortation, because I foresaw that the end of such a war would be most unfortunate to us. But I could not persuade them; for the madness of desperate men was quite too hard for me.[1]

Of course, Josephus realized that in arguing against the rebellion he was provoking the hostility and vengeance of the populace, who indeed might regard him as a traitor.

> I was then afraid, lest by inculcating these things so often, I should incur their hatred and their suspicions, as if I were of our enemy's party, and should run into the danger of being seized by them and slain.[2]

[1] Josephus: Life, 4.
Life, 5.

We quote Josephus here not as an individual, but as a representative of a type, to clarify to ourselves the attitude of Jesus. For on the surface may it not seem that Jesus held the same view as Josephus? Jesus, of course, opposed resistance to Rome. Hence does it not seem that they were in agreement toward the all-absorbing problem of the time? It may seem so, but it is not true. If it were true, nothing would have happened. Nothing happened when Josephus was speaking or writing. His writings are a matter of indifference to us and of no consequence. If Jesus had been thinking like Josephus there would have been no teachings of Jesus.

Those who favored non-resistance to Rome could be divided into two main types. One type welcomed and aspired to the universal Roman civilization. Complete assimilation, Greco-Roman culture was their ideal. Jewish national exclusiveness to them was nothing but provincial backwardness. They were an inevitable upper-class provincial phenomenon in the universalization of Rome and the Hellenization of the ancient world. To be a gentleman meant to them to be a Roman. In their hearts they accepted Rome. Their attitude towards religion was, of course, purely formal. There was, therefore, no occasion for any struggle whatsoever. Such a type could, of course, be neither numerous nor influential, but it undoubtedly existed. Rome could not expand politically without universalizing its own civilization, and such assimilation naturally appealed first of all to the upper class.

The other type of non-resistant was undoubtedly

numerous and significant. These were men who knew enough about the world at large to see clearly what resistance to Rome implied and foreboded. They knew that resistance was a physical impossibility and only invited complete destruction and devastation. They did not love Rome because they could not fight; they hated her the more. Their non-resistance was with a glowing eye and a heart full of hate, but with an arm that did not dare to strike. It was a prudent and practical attitude enough, but under the given circumstances it could not stem the tide. Sooner or later it was certain to be swept away by the tide of active resistance.

It could not stem the tide of brave, exalted resistance; it could not still the storm and allay the rising waves, because inwardly it shared their fury. It had no remedy against war, for it was itself latent war, counselling prudence. Prudence—is it really so prudent? Expediency—is it really so practical? Was it a livable life that prudence and expediency were dictating? They were counselling and preparing a life of outward submission and inward rage, a cringing life of stinging defeat with an inevitable outbreak at the end, when the accumulated burden of resentment should become unbearable.

It was all what we call very natural; in other words the solutions and alternatives, whether of rebellion or submission, were of the kind that float on the surface, that are obvious. They were in complete conformity with the age-long popular way of thinking and feeling. They offered but a stereotyped choice, and neither alternative contained a single new reaction, whether of thought or of feeling. These reactions, these ideas were so different from those which Jesus taught, that to teach as he did, Jesus must have had quite different inner reactions and experiences. Differing reactions, differing experiences, differing thoughts we reach and obtain only in a life that differs from the ordinary ease with its easy conclusions; in other words, in a life of inner struggle. Only in struggle life lifts itself out from the inherited and habitual grooves of feeling and of thought. In this struggle of Jesus' life, extraordinary insights and unique discoveries were reached, which in a fragmentary way are revealed to us in the gospels.

The problem that led to that inner struggle was neither secret nor precious; it was shouted from the housetops. But Jesus' solution, unlike the solution of Josephus, was unique. Historically considered, the problem was very local. Even from a religious point of view it was a provincial problem; yet Jesus' solution became the most universal achievement in the annals of mankind.

What the problem was historically speaking we know. But how does it present itself to an individual? It presents itself in the form of alternatives. I can not help feeling that the temptations of Jesus are probably parables of alternatives, of political and religious choices. Under this interpretation all the common popular solutions looked to Jesus like temptations of the Devil.

One solution could be expressed something like this:

Here is the holy city; here is the temple of God; and here are God's chosen people—His very own. Can God allow them to perish? Certainly not. Hence even the combat with the entire world, whose name is Rome, can not end but with the victory of God's own and only people.

> Cast thyself down: for it is written, He shall give his angels charge concerning thee: and in their hands they shall bear thee up, lest at any time thou dash thy foot against a stone. Jesus said unto him . . . Thou shalt not tempt the Lord thy God.[1]

Jesus did not accept the Zealot nationalist solution. There was, of course, an alternative in exactly the opposite direction: to let the Roman civilization supersede Judaism. Let the Jews frankly accept Rome and its culture, let them become Romans; then indeed the entire world will be theirs, and the glory of the Romans will be theirs as well.

> Again, the devil taketh him up into an exceeding high mountain, and sheweth him all the kingdoms of the world, and the glory of them; and saith unto him, All these things will I give thee, if thou wilt fall down and worship me. Then Jesus saith unto him, Get thee hence, Satan: for it is written, Thou shalt worship the Lord thy God and him only shalt thou serve.[2]

Between these two extreme solutions were, of course, many intermediary positions, chief among them the one that had no other aspiration than to live, and to live by bread alone. Such a solution neither sought nor required any religious sanction. But Jesus did.

Jesus was against resistance to Rome; but did he teach that it is expedient to submit, even with hatred in one's heart? Or did he teach:

[1] Matthew 4:6-7. [2] Matthew 4:8-10.

> Ye have heard that it hath been said, Thou shalt love thy neighbor, and hate thine enemy. But I say unto you, love your enemies, bless them that curse you, do good to them that hate you, and pray for them which despitefully use you and persecute you; That ye may be the children of your Father which is in heaven: for he maketh his sun rise on the evil and on the good, and sendeth rain on the just and on the unjust.[1]

Here indeed is quite a different solution of the problem, a solution that came to Jesus on fiery wings of exaltation.

The solution of every problem has some starting point. The starting point for Jesus was clearly the all-absorbing problem of the time. Jesus originally either resented the aggression of Rome, or he did not. If he did not, there was no occasion for any inner exertion. If he did resent, if he felt bitterly about it, what was he to do with himself and his resentment in this crisis? How could a proud spirit justify non-resistance to Rome? A proud spirit could not. But when the proud spirit exhausted itself in the struggle, came humility and acceptance, and with it exaltation embracing heaven and earth. The veil had fallen from the eyes, the simple meaning of the hidden things was revealed, and a new insight was won. With the certainty that only inner experience gives, Jesus could now show the way to the lost sheep of Israel.

> Come unto me, all ye that labor and are heavy laden, and I will give you rest. Take my yoke upon you, and learn of me; for I am meek and lowly in heart; and ye shall find rest unto your souls. For my yoke is easy and my burden is light.[2]

[1] Matthew 5:43-45.
[2] Matthew 11:28-30.

CHAPTER VII

SEVERAL years before the birth of Christ the Jews petitioned Rome for annexation to the Roman province of Syria, which petition Rome at that time rejected. The petitioners preferred to lose their quasi political independence under a Herodian prince for the sake of maintaining their religious traditions, for the sake of securing a cultural home-rule under a Roman procurator. The fifty Jewish ambassadors and the eight thousand Jews in the city of Rome who petitioned the Emperor were meeting a practical situation. This situation forced them to discount political independence altogether, and hence all that was left to save and safeguard was what had maintained them as a cultural entity throughout the ages—the traditional faith of their fathers.

A generation passed, and a similar problem, a similar alternative presented itself to Jesus; similar, but not identical, for the fullness of time was at hand; and on a wide and crowded road the children of Israel were rushing headlong toward their own perdition. The loud nationalist call to rebellion, the fervent hope for a Messiah, God's anointed leader and the redeemer of Israel, stirred the deepest emotions that human breasts could hold. Here was not a time for greater prudence, the time had come for the greater passions.

When Christ in ecstatic humility accepted submission to Rome, with that acceptance went as compensation the highest conceivable type of national consolation. Of course, it was not a material consolation for which the multitudes were looking, the great consolation was of a spiritual nature. The essence of this consolation was not wholly a stranger in the ideological literature of the nation. In the non-canonic literature of the time one finds many intimations of the universal mission of Zion; and in the canonic scriptures the noblest expression of this idea in the famous lines of Isaiah:

> And it shall come to pass in the last days, that the mountain of the Lord's house shall be established in the top of the mountains, and shall be exalted above the hills; and all nations shall flow unto it. And many people shall go and say, Come ye, and let us go up to the mountain of the Lord, to the house of the God of Jacob; and he will teach us of his ways, and we will walk in his paths: for out of Zion shall go forth the law, and the word of the Lord from Jerusalem. And he shall judge among the nations, and shall rebuke many people; and they shall beat their swords into plowshares, and their spears into pruninghooks: nation shall not lift up sword against nation, neither shall they learn war any more.[1]

Of course, Isaiah's Zion is judging, Jesus' Zion is saving. Still in Isaiah is an indication of Christ's consolation for the children of Israel. Though they were losing their political independence, how trifling it is in the light of their universal calling. They were indeed to be God's chosen people, God's light in a world of darkness; "for salvation is of the Jews." [2]

[1] Isaiah 2:2-4.
[2] John 4:22.

Now let us turn to the gospels, to the opening passage in the Sermon on the Mount, and consider it not from a religious but from a historical viewpoint. It begins with blessings upon the humble: "Blessed are the poor in spirit; for theirs is the kingdom of heaven. Blessed are they that mourn:" (I dare say one mourns the loss of one's national independence) "for they shall be comforted. Blessed are the meek: for they shall inherit the earth." [1] Of course, humility, mourning (which means accepting the will of God as Job did; the Jews still in personal mourning recite the book of Job) meekness, hungering and thirsting after righteousness, are all spiritual terms; and to inherit the earth means but a spiritual inheritance. Therein is the consolation. This is further clarified in the passage beginning, "Ye are the salt of the earth." Clearly it is not addressed to the world at large, for then there would have been no earth left, only salt. It might have been said in other words, You are the chosen people, but for what were you chosen? Chosen to carry to the world a spiritual message. If you have no spiritual message for the world what are you good for?

> But if the salt have lost his savor, wherewith shall it [the earth] be salted? It is thenceforth good for nothing, but to be cast out, and to be trodden under foot of men.[2]

The same idea only with different imagery is carried out in the following verses:

> Ye are the light of the world. A city that is set on an hill cannot be hid. Neither do men light a candle, and put it under a

[1] Matthew 5:3-5.
[2] Matthew 5:13.

bushel, but on a candlestick; and it giveth light unto all that are in the house. Let your light so shine before men, that they may see your good works, and glorify your Father which is in heaven.[1]

Christ says your Father, not their father, because he is addressing the chosen people as the children of God, who have a spiritual mission to perform.

So far as the law and prophets are concerned there is no infringement upon them in the Sermon on the Mount. It was not less piety or less righteousness that Jesus preached.

Think not that I am come to destroy the law, or the prophets: I am not come to destroy, but to fulfill.[2]

Why then do we find in the Beatitudes this passage?

Blessed are they which are persecuted for righteousness' sake: for theirs is the kingdom of heaven. Blessed are ye, when men shall revile you, and persecute you, and shall say all manner of evil against you falsely, for my sake.[3]

Are the public hatred and persecution here referred to due to contemporary religious bigotry, religious intolerance, which could not listen to a more spiritual interpretation without violence? Were the people and their leaders so intolerant of greater religious fervor or greater liberalism than their own little minds by chance were capable of? What little we know about actual conditions and circumstances of the time would hardly support such a view. The circumstances forced a very unusual degree of religious toleration.

First of all, the Sadducees and Pharisees had to learn to get along and worship in the same temple.

[1] Matthew 5:14-16.
[2] Matthew 5:17.
[3] Matthew 5:10-11.

Secondly, nobody seemed to disturb the "sinners," that is, the outright and outspoken religious liberals. The differences between the Sadducees and the Pharisees must have been tremendous. We are told that the Talmud places the Sadducees on a level with the Samaritans.[1] The Sadducees did not accept the rabbinic traditional interpretations of the Bible. The Pharisees, on the other hand, regarded it as "more culpable to teach contrary to the precepts of the scribes than contrary to the Torah itself." [2] Still more drastic is the difference between the sects in their attitude toward immortality. The idea of resurrection or immortality of the soul was completely rejected by the Sadducees. Great as were their differences of viewpoint, these varying sects did not persecute each other.

Why then should Jesus assume that his followers will be reviled and persecuted? Is it because of the Christhood of Jesus? But did not his own people in Nazareth try to kill him before he acknowledged his Christhood? One does not need to look very far to find the reason for the antagonism to Jesus. Was it not he who in the midst of the brewing rebellion was teaching:

> That ye resist not evil: but whosoever shall smite thee on thy right cheek, turn to him the other also.[3]

[1] Nidda IV, 2. "The daughters of the Zaddukim are, if they walk in the ways of their fathers, equal to Samaritan women. If they walk openly in the ways of Israel, they are equal to Israelitish women. R. Joses says: They are looked upon as Israelitish women, unless it is proved that they walk in the ways of their fathers."—Quoted in Schürer's History of the Jewish People in the Time of Jesus Christ. Second Division, v. 2 (English translation), p. 8.

[2] Sanhedrin 11, 3.—Schürer, *Ibid.*, p. 12.

[3] Matthew 5:39.

It was Jesus who was teaching:

> But I say unto you, Love your enemies, bless them that curse you, do good to them that hate you, and pray for them which despitefully use you, and persecute you.[1]

Under the circumstances, therefore, those who understood and followed Jesus were certain of meeting violent antagonism from a people that was on the eve of rebellion and disaster. Political passions were, of course, clothed in traditional religious terms. Messianic hopes in no wise changed the tribal traditional morality of the people: such hopes rather enhanced it. What then could save the people? Only that great spiritual experience, the passionate and humble submission to the will of God; only a rebirth in spirit could save them from their traditional reactions. Without this new glowing spirit, the old tribal morality, the standards of flesh were sure to prevail.

> Verily, verily, I say unto thee, Except a man be born again, he cannot see the kingdom of God.[2]

The Zealot movement as such was of relatively recent origin, but was linked to the most ancient traditions: tribal morality and religious orthodoxy. Orthodoxy, on the other hand, no matter how genuinely devout and pious it may be, is in its very nature a historical, inherited, traditional formulation and observance. In the historical moment that we are dealing with only a religious fervor of so passionate a nature that it could overcome traditions and habits and all the emotions aroused by the day, only such fervor could

[1] Matthew 5:44.
[2] John 3:3.

save the people from perdition. The call was, therefore, for a greater ruling religious passion, a passion of which clearly not everybody was capable. Only a part of the nation at best could free itself from the traditional nationalistic reactions, from the traditional habits and the traditional viewpoint.

> For I am come to set a man at variance against his father, and the daughter against her mother, and the daughter in law against her mother in law. And a man's foes shall be they of his own household.[1]

Did the Zealots ever try to save their lives? For the God of their fathers and the freedom of their country they would unflinchingly sacrifice not only their own lives, but the lives of all who were dear to them. What doubt could there be how they were bound to view the teaching of Christ even if their own brother, their own child should profess it?

> And the brother shall deliver up the brother to death, and the father the child; and the children shall rise up against their parents, and cause them to be put to death.[2]

He to whom the ties of life, the ties of old were too strong—he really could not be Jesus' disciple. Hence the extraordinary text:

> If any man come to me, and hate not his father, and mother, and wife, and children, and brethren, and sisters, yea, and his own life also, he cannot be my disciple.[3]

It was the call for a religious revival. The very call for repentance was nothing else. Love God, your Father, with all your heart and all your soul. Submit

[1] Matthew 10:35-36.
[2] Matthew 10:21.
[3] Luke 14:26.

gladly to His will. Pray that not your will but His be done here on earth as in heaven. That is all. This simple call for a spiritual revival was, however, offensive to the prevailing political sentiment as well as to organized religion. The inevitable situation developed. All worship of God is the product of a religious organization with its teachings, formulas, observances, rituals, and traditions of the elders. An organization, even to maintain a spiritual entity, is in its very nature a physical, material instrument, whose object is to provide the many with at least a minimum of spiritual ity. But when the day came when not a minimum but a maximum of possible human spirituality was called for, then indeed all the traditional trappings of the old organization conflicted with the very object for which they had been created. The vehicles of a religion were too heavy to permit any soaring of the spirit. Yet without such new spiritual content, without a newly felt relationship to their heavenly father, without a universal mission, there was no consolation left for those who were about to perish.

CHAPTER VIII

WERE the reactions of so unique a religious personality only emotional, or did Jesus have also a unique intellectual insight? There is no question in my mind that Christ's deep conviction that his is the Way and the Truth was based on knowledge, intellectual knowledge, scientific knowledge if you please. Before he felt that he was the Redeemer, he knew himself to be the great Discoverer. Of course, this is a modern mode of expression. We in the twentieth century talk and think of *our* discoveries, of *our* personal achievements; but to Jesus a concrete and self-evident intellectual insight was a gift of God. Truth could only come from the source of all truth; from the Father that is in heaven.

Is the complete revelation of Christ's intellectual discoveries in the gospels? Could it be there? What are the gospels? At a certain time, Christ taught. Multitudes were gathered around. He talked to them and answered questions. His sayings on these occasions were remembered, sometimes possibly verbally, sometimes inaccurately. At different times after Christ, these sayings were gathered and edited. To them were added records of his deeds, of his healing, and other material which human memory and tradition associated with Christ. Christ did not write a philosophical treatise about his knowledge of life and of God.

Take any contemporary example: Let us assume that a great Christian philosopher and thinker, for example Tolstoi, on the basis of his insight into what he considers truth, is trying to teach his fellow men, as Tolstoi actually did. But now let us assume that all his literary and religious writings were not written, but that the only records left to us were his pedagogical efforts, his little tales and stories for the peasants (which, incidentally, I believe have never been translated out of Russian). All his little tales for the education and spiritual uplift of the peasant are based on a rather profound and complex intellectual insight. But you could hardly expect to find dissertations on philosophy in stories written for the poor, ignorant peasants of Russia. Yet I venture to say, were all the works of Tolstoi destroyed and only these simple folk-tales left, that it would take a very naïve scholar not to see the intellectual and religious system that lies behind these tales written for poor, degraded toilers of the soil.

This is a hypothetical example. We are actually infinitely better off with the gospels. True, they are largely teachings of conduct. True, they are sayings addressed to men and women from whom much could not be expected intellectually. True, there is no attempt at a philosophical and theological dissertation; and yet there was no need for followers of Christ to go to an Aristotle for philosophy. For a greater than Aristotle is there in the very sayings as they have been recorded and have come down to us in the gospels.

What is a philosopher? A lover of wisdom it means philologically. And what is wisdom? A relative insight into truth, very relative indeed. What then shall we call Christ, who knew that he had not a relative but an absolute insight? Moreover, use all your modern little scientific standards, and you cannot get away from the fact that Christ's insight was one which future generations may rediscover but can never upset.

Is it, therefore, surprising that Christ knew quite well that he was wiser than Solomon?

> The queen of the south shall rise up in the judgment with this generation, and shall condemn it: for she came from the uttermost parts of the earth to hear the wisdom of Solomon; and, behold, a greater than Solomon is here.[1]

What that great revelation was we will discuss presently. And it is for this insight that the great thanksgiving was rendered by Christ, probably approximately expressed in Luke:

> I thank thee, O Father, Lord of heaven and earth, that thou has hid these things from the wise and prudent, and hast revealed them unto babes: even so, Father; for so it seemed good in thy sight.[2] . . . For I tell you, that many prophets and kings have desired to see those things which ye see, and have not seen them; and to hear those things which ye hear, and have not heard them.[3]

It is because back of all the teaching was an insight that carried with it complete conviction of self-evident truth that Jesus taught "as one having authority, and not as the scribes,"[4] and that Jesus could say to Nicodemus:

[1] Matthew 12:42.
[2] Luke 10: 21.
[3] Luke 10:24.
[4] Matthew 7:29.

Verily, verily, I say unto thee, We speak that we do know, and testify that we have seen; and ye receive not our witness.[1]

What is it that Jesus knew, and what is it that he had seen in his own experience, that was hidden from the kings and prophets? It is condensed in a very brief formula—The kingdom of heaven is in us. This formulation, however, may be likened to the summit of a mountain. The entire broad base, the vast expanse of the mountain's height and breadth support and lead up to the peak. That mountain peak is but the crowning glory of the mountain's vastness, a vastness of insight based on experience. Christ was speaking of what he knew, of what he had seen. What did he see, what did he experience? All that he experienced we do not know; but an outline here and there suggests its depth and indicates its bulk.

From our historical analysis of the situation it becomes quite evident that Jesus had to resent deeply the loss of Jewish national independence and the aggression of Rome. Had he not resented it there would have been no cause for his fervent humility and acceptance. The fervor and ultimate depth of the reconciliation leave under the given historical circumstances no doubt as to the character of the struggle which preceded it. What happened? National humiliation was hurting and burning. The balm for that burning humiliation was humility. For humility cannot be humiliated. Did humility change the outside world? Not in the least. Only an inward change took place; yet that inward change completely altered the

[1] John 3:11.

so-called facts of life and of existence. Thus he asked his people to learn from him,

For I am meek and lowly in heart: and ye shall find rest unto your souls. For my yoke is easy, and my burden is light.[1]

Parallel with the great emotion was the intellectual insight, that what counts in life and constitutes life is the inner reaction; and that so-called outward facts to which we have no inner reactions are not part of our life. The outward world is our world only in so far as *we* react to it. Great may be the bulk of yonder distant star; and in the scheme of the universe its significance may be greater than that of our little planet. But in our life its bulk and gravity count for little; for to our reactions it is but one of innumerable little stars which in no wise affect our lives. The same, of course, is true about things nearer home. In so far as we do not react towards some phenomena of life those phenomena do not exist for us. It is our reaction, our attitude that so far as we are concerned gives to any phenomenon its place and meaning in our life. It is, therefore, with the inner attitude which determines our reactions and thus regulates all the events of our lives that Christ was dealing. Christ was illustrating this viewpoint of his when he said:

The light of the body is the eye: if therefore thine eye be single, thy whole body shall be full of light. But if thine eye be evil, thy whole body shall be full of darkness. If therefore the light that is in thee be darkness, how great is that darkness![2]

You can see that Christ is fully conscious of this principle, and expresses it, we may say formulates it

[1] Matthew 11:29-30.
[2] Matthew 6:22-23.

intellectually, in connection with many cures which are reported in the gospels. I am talking about the miraculous cures. Christ's attitude toward miracles in general can be seen in the so-called temptations in the desert. You know what he thought of the Pharisees when they asked him for a sign in heaven. He considered a generation that wants a sign "a wicked and adulterous generation"; for, of course, all such signs would have been outward forces and hence meaningless. On the other hand, see what he says to the people who come to him afflicted with bodily ills. We have a statement in Matthew:

> And when he was come into the house, the blind men came to him: and Jesus saith unto them, Believe ye that I am able to do this? They said unto him, Yea, Lord. Then touched he their eyes, saying, According to your faith be it unto you.[1]

Here are two blind men praying him to heal them. He asks them whether they think he can do so; that is, whether they have such inner faith. Then all that he tells them is, "According to your faith be it unto you." In the same chapter you will find a woman who insisted on touching the garment of Christ. She was cured of her ailment. She did not even ask Christ to cure her, but she had an inner faith. Christ became aware of the cure only *post facto*. He tells her, "Thy faith hath made thee whole."[2] On another occasion he is again confronted by a blind man. Again he cures him and says to him, "Go thy way; thy faith hath made thee whole."[3] A similar formula you will find in other cases.[4]

[1] Matthew 9:28-29.
[2] Matthew 9:22.
[3] Mark 10:52.
[4] Luke 17:19, 18:42.

We also know, on the other hand, that in many cities, like Nazareth and other places where people did not believe in Jesus, he could not perform any miracles. These miracles, therefore, and Christ was quite conscious of it, were acts of faith, inner acts of the afflicted. True enough, the children of Israel, without any faith in him, in fact with doubt instead of faith, wanted from him some miraculous signs in heaven to prove his Christhood. Such signs, of course, were not given.

The record of one of these cures links the cure with forgiveness of sin, which is intellectually very interesting and exciting. It shows how highly systematized was Christ's intellectual insight. It cannot be a mere chance interpolation of the editor of the gospels. Do you remember the case of the man sick of the palsy? [1] This man had faith in Christ and wanted to be cured. Christ says to him, "Thy sins are forgiven," and the Pharisees wonder who the man can be who has power to forgive sins. But Christ identifies his healing and his forgiving sins in the statement, "Whether it is easier, to say, Thy sins be forgiven thee; or to say, Rise up and walk?" He identifies the two acts because

[1] Luke 5:18-23. And, behold, men brought in a bed a man which was taken with a palsy: and they sought means to bring him in, and to lay him before him. And when they could not find by what way they might bring him in because of the multitude, they went upon the housetop, and let him down through the tiling with his couch into the midst before Jesus. And when he saw their faith, he said unto him, Man, thy sins are forgiven thee. And the scribes and the Pharisees began to reason, saying, Who is this which speaketh blasphemies? Who can forgive sins, but God alone? But when Jesus perceived their thoughts, he answering said unto them, What reason ye in your hearts? Whether it is easier, to say, Thy sins be forgiven thee; or to say, Rise up and walk?

both the cure and the forgiveness of sin are made possible by the inner act of the man himself. This becomes even more evident in the case of the woman who loved much. Here are the verses:

> And he said unto her, Thy sins are forgiven. And they that sat at meat with him began to say within themselves, Who is this that forgiveth sins also? And he said to the woman, Thy faith hath saved thee; go in peace.[1]

Now, perhaps, we understand why Christ tried to explain and elucidate his own acts to the scribes and Pharisees by the acts of John the Baptist.

> The baptism of John, whence was it? from heaven, or of men? And they reasoned with themselves, saying, If we shall say, From heaven; he will say unto us, Why did ye not then believe him?[2]

The three synoptic gospels have obviously one source for the record of this conversation, and the wording of this source seems rather ambiguous. It looks as if Christ refused to explain to the Pharisees the character of his authority, or tried to put the Pharisees in the difficult position of having either to accept or deny the authority of John. As a matter of fact, that conversation is marvellous in its explicitness, and again it shows how systematic and thought out is the insight of Christ. What was John the Baptist doing? He denied that he was Christ, he did not even acknowledge himself as a prophet. He called himself "a voice crying in the wilderness," and described his mission as to make the path straight for him that was to come. Yet John was remitting sins, for baptism was a token of the remission of sins. Now, did John discriminate among the people who came to him to be baptized?

[1] Luke 7:48-50

[2] Matthew 21:25.

Did he refuse baptism or remission of sin to any one? He did not. If they repented and changed their attitude and were yearning for remission of their sins, they were baptized. A change of mind has already taken place; in their repentance was the remission of their sins. Publicans and harlots repented and their sins were forgiven, the baptism was but a token thereof.

It was difficult for the Pharisees to understand it. Religion to them was largely a matter of outward regulation, the ultimate significance of the inner attitude was incomprehensible to them. If they lived up to all their religious regulations they had consciousness but of their piety and righteousness. They had no yearning for spiritual rebirth, and nothing could be done for them. They were cleansing and polishing the outside of the cup.

We know that the Jews expected God to send their deliverer, and expected that with him a new rule would begin, by a ruler sent from God himself—the kingdom of heaven. So far, therefore, as the masses are concerned, the deliverance of the Jews and the kingdom of heaven were acts of God, external acts. I daresay they would have expected the kingdom to be inaugurated by signal victories over the Gentiles, by God's judgment and chastisement of publicans and sinners, and what not.

All this from the viewpoint of Christ's intellectual insight was futile nonsense; for no external act could solve this or any other situation. One could not enter into the kingdom of heaven without a rebirth in spirit.

It was only through a rebirth in spirit that one could enter therein. The kingdom of heaven was but an inner change in us. True enough, the inner spiritual change may be gradual. It may be like the plant that grows from a tiny mustard seed. It may be like a leaven, which raises the loaf gradually. But the leavening and the growth are inner acts, not outward manifestations. Was it not a completely different concept from the one which then prevailed? Indeed it was. And that is why Christ told the Jews that they knew neither the Father nor the Son:

> Ye neither know me nor my Father: if ye had known me, ye should have known my Father also.[1]

According to the popular conception, Christ was to inaugurate the kingdom of heaven. Christ was to save the lost sheep of Israel, to save them in the last moment from impending destruction. Now Jesus knew quite well that his way was not simply one way to save the children of Israel, but the one and only way. That way was to instruct them in the kingdom of heaven. Thus Christhood, the kingdom of heaven, and the salvation of Israel remain linked together, as in the popular concept. But in Jesus' concept there appears this difference: that Christhood and the salvation of Israel and the kingdom of heaven postulated the spiritual rebirth of the people.

> And when he was demanded of the Pharisees, when the kingdom of God should come, he answered them and said, The kingdom

[1] John 8:19.

of God cometh not with observation: Neither shall they say, Lo here! or, lo there! for, behold, the kingdom of God is within you.[1]

But the Pharisees could neither enter into that kingdom themselves nor could they suffer others to enter therein.

[1] Luke 17:20-21.

CHAPTER IX

ACCORDING to Josephus, John the Baptist was put to death for purely political reasons. He tells us:

> Now when many came in crowds about him, for they were greatly moved by hearing his words, Herod, who feared lest the great influence John had over the people might put it into his power and inclination to raise a rebellion (for they seemed ready to do anything he should advise) thought it best, by putting him to death, to prevent any mischief he might cause; and not bring himself into difficulties by sparing a man who might make him repent of it when it should be too late.[1]

John the Baptist, therefore, according to the very plausible testimony of Josephus, was put to death for political reasons. What did John the Baptist do? He announced the coming of the Messiah. The Messiah, in the general and universal understanding of the time, was to be the deliverer of the Jews from Roman oppression. Herod, who had received his appointment as tetrarch of Galilee from Rome, was but an administrative instrument of his Roman sovereign. To him the coming of Christ could mean nothing but rebellion against Rome, under a leadership which the people would acclaim as divine. Whatever may have been the flavor of John's religious and moral preachings, to Herod he was but the herald of a revolution, with great moral power over the people, who came to him in multitudes. Since the fate of the Herodians was

[1] Josephus: Antiquities, XVIII, 5, 2.

tied up with the power of Rome, Herod put the precursor of what looked to him like the coming revolution, to death. To the Jewish populace, the Christ was the deliverer who was to come to deliver them from foreign rule and oppression. To a Herod or a Pilate, or any Roman administrative agent, the Christ who was to come was the leader of the expected rebellion. For what looked to the so-called Jewish patriot like deliverance, of course, meant rebellion to the forces of Rome.

John was put to death by Herod for announcing the coming of the Christ. Yet when Pontius Pilate sent that very Christ to Herod, Herod did not put him to death, but sent him back to Pilate; and neither Pilate nor Herod could find any fault with him. The Jews, on the other hand, who were praying so fervently for the coming of the Christ, sought from Pilate Jesus' execution and the deliverance of the rebel leader, Barabbas.

These historical episodes throw light on the wide gulf between the two concepts of Christhood; that of the populace, and that of Jesus. The concept of the populace was a heavensent king of the house of David, with a supernatural sword in his hand, ruling, judging and avenging. Very different was Jesus' concept of his Christhood.

A very large number of the plain people believed in Jesus. They saw before them a personality whose like they had never seen before. They believed him to be the one who was to come; that is, the Christ that they expected, whose functions and attributes were those popularly attributed to the coming Messiah. Even Christ's own disciple, Simon Peter, who according to the gospels first acknowledged him to be the Christ, even he seems to have understood Christhood in quite a different sense from that of Jesus, and, in all probability, in the ordinary sense of the people. The way Christ took it is described:

> But he turned, and said unto Peter, Get thee behind me, Satan: thou art an offense unto me: for thou savorest not the things that be of God, but those that be of men.[1]

The popular interpretation of Christhood and the popular faith in Jesus as the Messiah are also indicated by Jesus' having to hide himself in a mountain lest the populace should by force make him king.

> When Jesus therefore perceived that they would come and take him by force, to make him a king, he departed again into a mountain himself alone.[2]

There is little doubt that so great an impression was made by Jesus upon his day and generation that had he wanted to be a king and lead his people as the Messiah they expected should have done, he would have been joyfully acclaimed throughout Judea. Yet Jesus chose and had to choose the cross. Why did he have to reject the throne? Why did he have to choose a cross? Why did he have to change the definite meaning that so concrete a word as Christ or Messiah had in his time? It could be answered by quoting John:

> I am come a light into the world, that whosoever believeth on me should not abide in darkness.[3]

[1] Matthew 16:23.
[2] John 6:15.
[3] John 12:46.

This is an answer; but let me give one less subtle, more concrete.

It has been indicated in the foreword to this study, I believe, that our attempt here is neither theological nor philosophical, but an attempt at historical understanding. Understanding, however, is not a vain repetition of many words and enumeration of various parts of something. It is an understanding of the inner cohesion of these parts, that gives them an entity and intellectual unity. To me personally it seems childish not to see in Christ's teachings an overwhelming intellectual system. The towering parts that are its components are parts of the same system, not independent units. The truth of the insight, the cohesion of the system were self-evident to Christ; so much so that he knew that they had an absolute quality; that is, coming from God. Because of the systematic nature of the insight, the conclusions drawn were inevitable and mandatory.

Just so inevitable was a revision of the concept of Christhood. Had all the popular functions of Christhood been excluded from Jesus' concept, then indeed Jesus would have simply brushed away the entire concept. He would have said, "No, I am not he that is to come," or "He will come," or he would have said, "Indeed he never will come." But the primary and elementary function of the Christ that was to come was what? The saving of the Jews, was it not? Jesus knew quite well that the only thing that could possibly save them was his insight, as expressed in his teachings. He, therefore, completely fulfilled the fundamental meaning of Christhood. No one but himself, moreover, could possibly fulfill it. That he considered it his primary function is shown in the way he commanded his disciples:

> These twelve Jesus sent forth, and commanded them, saying, Go not into the way of the Gentiles, and into any city of the Samaritans enter ye not: But go rather to the lost sheep of the house of Israel.[1]

While he was instructing his people into the kingdom of heaven, leading them to a spiritual rebirth, Jesus nevertheless kept constantly before them the pragmatic importance of his teachings, which could save them from imminent destruction. He does not hesitate to show the plain people that the very political and social situation, that is, the times in which they were living, demanded from them a changed attitude of mind unless they were to perish; though, of course, it would have been preferable that they change their attitude, not because of the existing political situation, but because it was right that they should. So for instance we find in Luke these sayings:

> And he said also to the people, When ye see a cloud rise out of the west, straightway ye say, There cometh a shower; and so it is. And when ye see the south wind blow, ye say, There will be heat; and it cometh to pass. Ye hypocrites, ye can discern the face of the sky and of the earth; but how is it that ye do not discern this time? Yea, and why even of yourselves judge ye not what is right?[2]

But they really could not judge what was right, for their minds were filled with the conglomeration of

[1] Matthew 10: 5-6.
[2] Luke 12:54-57.

popular ideas in which the coming of Christ and his kingdom, the salvation of Israel, the tribute to Cæsar, and endless other religious and secular ideas of the time were all mixed up and intertwined.

The popular current concepts presented a curious mixture of things religious and things political, of things natural and supernatural. They were products of an emotional panic which was hysterically fusing and confusing things. In the Messianic, apocalyptic and eschatological literature of the time the world was to come to an end; but what really did come to an end in that literature was the last shred of thinking capacity and common sense. In Christ, on the other hand, in his teachings, his ministry, entirely apart from any of his other functions or qualities, the one thing that stands out monumentally is his intellectual grandeur, and the purity and unswerving consistency of its simple straight lines. The continuation of a straight line excludes doubt as to its direction. The line of Jesus' intellectual insight had to lead to a recasting of the concept of Christhood, no matter how widely the concept he arrived at might vary from the confused and uncertain one which prevailed.

The two salient points common and fundamental to all prevailing concepts of the Messiah were the salvation of Israel and the inauguration of the kingdom. Since the kingdom could only be within the souls of men, since salvation of Israel from immediate destruction was dependent on the humility and nonresistance which would accompany a spiritual rebirth, Jesus knew that he was *the* Christ, and that any other

Christ who might arise, a Christ who would be a popular leader, was bound to be a false Christ. It was equally self-evident that the Messiah of the popular imagination, the man on horseback, of the conquering hero type, could accomplish nothing but destruction. He could accomplish nothing because the only conquest required for entrance into the kingdom of heaven was an inner conquest. Even if material conflict should be crowned with victory, what would such victories of the flesh avail? How could the Messiah of the popular imagination lead the Jews to a rebirth of the spirit, and to the gates of the kingdom that is within us and that cometh not with observation? How could men possibly enter into the kingdom, supposing some external changes to take place, if they themselves remained unchanged? But if our entrance into the kingdom is entirely a matter of changing our own attitude, of our own rebirth, what else could Christ be but a light to those that sit in darkness, and their minister?

"Ye know that they which are accounted to rule over the Gentiles exercise lordship over them; and their great ones exercise authority upon them. But so shall it not be among you: but whosoever will be great among you, shall be your minister: And whosoever of you will be the chiefest, shall be servant of all. For even the Son of man came not to be ministered unto, but to minister, and to give his life a ransom for many.[1]

All the ideas of Jesus were correlated; they were closely fitted parts of one great intellectual concept, and all of the same spirit, a different spirit from the one prevailing.

[1] Mark 10:42-45.

There can be no doubt that many, very many believed in him; but how many understood him? Certainly very few. The gospels themselves as they come to us testify to the lack of understanding even among the disciples. So we are told:

> And they understood none of these things: and this saying was hid from them, neither knew they the things which were spoken.[1]

The great trouble was that Christ was teaching an insight, preaching ideas, while the people could only understand things. So, for instance, even so simple a metaphor as "the leaven of the Pharisees and the leaven of Herod" was understood literally and materially.

> And they reasoned among themselves, saying, It is because we have no bread. And when Jesus knew it, he saith unto them, Why reason ye, because ye have no bread? perceive ye not yet, neither understand? have ye your heart yet hardened? Having eyes, see ye not? and having ears, hear ye not? and do ye not remember?[2]

You will find any number of references in the gospels to this lack of understanding. Some are even humorous; for Jesus, of course, could not help seeing that they were hopelessly mixing his teachings with the old traditional ideas.

> Therefore every scribe which is instructed unto the kingdom of heaven, is like unto a man that is a householder, which bringeth forth out of his treasure things new and old.[3]

The scribes had difficulty in grasping the meaning of Jesus' message; but was it any easier for those who were not scribes? To illustrate and illuminate his

[1] Luke 18:34.
[2] Mark 8:16-18.
[3] Matthew 13:52.

teachings Christ used parables, but they did not help very much.

> Therefore speak I to them in parables: because they seeing see not; and hearing they hear not, neither do they understand. And in them is fulfilled the prophecy of Esaias, which saith, By hearing ye shall hear, and shall not understand; and seeing ye shall see, and shall not perceive: For this people's heart is waxed gross, and their ears are dull of hearing, and their eyes they have closed; lest at any time they should see with their eyes, and hear with their ears, and should understand with their heart, and should be converted, and I should heal them.[1]

And yet there is no doubt that multitudes believed in him, and that faith did wonders for them. The more reason is it for us to ask ourselves, Why was not their faith abiding? Why was it that the multitudes who greeted him with "Hosanna," the very same, perhaps, cried but a few days later, "Crucify him"? Why was it that Jesus knew that he must be rejected by his generation and suffer many things in Jerusalem? And not only Jesus knew it, but his brethren, who did not believe in him, taunted him, and asked him why he was not going to Jerusalem.

> Then Jesus said unto them, My time is not yet come: but your time is always ready. The world cannot hate you; but me it hateth, because I testify of it, that the works thereof are evil. Go ye up unto this feast: I go not up yet unto this feast; for my time is not yet full come.[2]

Let us try to confront the situation, and find ourselves in it. Jesus might have been followed and grasped in one of two ways: first, his teachings might have been understood, believed in, and followed with

[1] Matthew 13:13-15.
[2] John 7:6-8.

abiding conviction because of that understanding. Certainly he laid emphasis on understanding, and pointed out that

> When any one heareth the word of the kingdom, *and understandeth it not,* then cometh the wicked one, and catcheth away that which was sown in his heart. This is he which received seed by the wayside.[1] . . . But he that received seed into the good ground is he that heareth the word, *and understandeth it;* which also beareth fruit, and bringeth forth; some a hundredfold, some sixty, some thirty.[2]

The alternative possibility is that the teachings might not have been intellectually understood, but that Jesus might have been felt and grasped emotionally, and followed because of the people's faith in him. Have we any definite evidence that Christ's unified intellectual insight was understood and mentally grasped? After nineteen hundred years of all kinds of theology, philosophy and science, we can understand it to-day. Whatever one may think of our intellectual achievements, be they profound or not, it is fair, I believe, to say that we at least can grasp an intellectual insight if it is laid before us; thus science and philosophy have really paved the way to an understanding of Christ. But I frankly fail to see how in Christ's generation, in Judea and by the shore of Galilee, there could be many who would understand him. The political emotional elements in the situation would also have worked against a sympathetic understanding, for all that the scribes could understand and did understand was that if men should truly be-

[1] Matthew 13:19.
[2] Matthew 13:23.

lieve in him, then "The Romans shall come and take away both our place and nation."[1] There was altogether too tremendous a difference in the intellectual level; and I cannot see how an intellectual understanding of Christ was at all possible.

Quite different is it with faith, and it may well be observed that so-called intellectual understanding does play a rôle only in the history of so-called ideas; that is, understanding plays a rôle in the history of understanding—a trite enough observation. It plays an infinitely small rôle in the history of mankind. Mankind and understanding are two different things. You perhaps witness from time to time great commotions in the name of ideas. It was so in the past, is perhaps so in the present. Do not think for a moment that it is understanding of the ideas which moves mankind; it is their faith in the ideas. This is true about the so-called masses, it is true about so-called intellectuals; when at certain times numbers of persons call themselves positivists, Kantians, Hegelians, Marxists, all you will find there is sincere and really powerful faith in the concepts of Comte or Kant or Hegel or Marx. That faith is clothed (because it is so scientific, because it is after the fall of Adam and no longer in a state of innocence) it has to be clothed in phrases and excerpts—rags of the believers' particular master. So it is with faith; and as it is, so it was, and so perhaps it will be. But there are certain fundamental conditions, subject not to faith but to understanding, that at a given time determine the general

[1] John 11:48.

characteristics of a prevailing faith. The faith that had to prevail in the generation of Jesus was a faith in Christ, *their* Christ, *their* Messiah. And they believed in Jesus.

But here was a fatal tragedy involved in that very faith; for one concept was the concept of the Messiah of Jesus' generation, and different, as we pointed out, was Jesus' concept of his own Christhood. If the insight of Jesus could have been intellectually grasped, they could and would inevitably have come to that concept of Christhood that Jesus taught. But the Messiah was too definite a concept of faith to be modified without a sign in heaven. Multitudes believed in Jesus; and the whole of Galilee and Judea would have been swept by Jesus, could he have been the king and the Messiah of their faith. For whenever they believed in him, they believed in him *as their* Messiah, their anointed king. "Hosanna, blessed is the king of Israel that cometh in the name of the Lord." [1] They believed in him, but they believed in him with *their* faith, not with *his* faith. How could they modify so deeply ingrained a concept of faith, and a concept of deliverance that the political situation seemed to them so concretely to demand? Where intellectual understanding was lacking, nothing short of a sign in heaven, an intervention by God himself, could modify their faith. And in this respect, if we come to think of it, we must remember that even the faith of the early Christian community after the death of Jesus was based first upon a sign in heaven, the resur-

[1] John 12:13.

rection. Nor was the current faith in the Messiah so drastically changed, for he was to come again and reign in glory. Even after the resurrection the "Acts" are reporting the old primary concern:

> When they therefore were come together, they asked of him, saying, Lord, wilt thou at this time restore again the kingdom to Israel? [1]

Jesus knew that he had to be rejected by his generation. And the Pharisees knew just how to shatter the faith in Jesus as the Messiah. For indeed all they had to ask him was the question whether it were lawful to pay tribute to Cæsar. The Messiah that was to deliver the children of Israel from the Cæsars and all oppression, that Messiah could not command them to pay the tribute. But Jesus, who came to deliver them from themselves and from their imminent destruction, of course, had to tell them to recognize the fact that on their tribute money were the name and the superscription of Cæsar, hence to render unto Cæsar what was Cæsar's, but to give unto God what was God's.

Jesus had to be rejected by his generation, and he knew it. If he was to be rejected by his generation and suffer for the truth to which he came to bear witness, then indeed he could not save the lost sheep of Israel from their imminent destruction. Rejected by his generation and not understood by his people, of what avail was his instruction into the kingdom? Will not the people to whom he ministered after all

[1] Acts 1:6.

be the very last to enter the kingdom, if indeed they are to enter therein at all? Will not his very disciples be offended and deny him, when he, instead of reigning in power, is apprehended like a malefactor and suffers at the hands of his enemies? And true enough, when he was apprehended as a malefactor, then the disciples forsook him and fled.[1] They were offended, they forsook him, for neither was his kingdom of this earth, nor was there any heavenly intervention in his behalf. If there had been, his enemies would have believed in him as well. Did not the priests and scribes say,

> Let Christ the King of Israel descend now from the cross, that we may see and believe.[2]

Thus Christ had to experience a greater passion than the physical one.

It is generally said that human passions are blind, blind to causes, conditions, consequences; blind, that is, having no insight into more general conditions of existence. And because it is so, "If the blind lead the blind, both shall fall into the ditch." [3] On the other hand, to every historical moment, transient as it is, its momentary passions are by far more absorbing and exciting than a general insight, if ever so true, into life. These passions of the moment have naturally enough their spokesmen. More universal viewpoints may also have their spokesmen. But in a conflict between the moment and eternity, which is it that is going immediately to conquer? Unquestionably the moment;

[1] Matthew 26: 56.
[2] Mark 15: 32.
[3] Matthew 15:14.

for it is the moment that is passionate, blind and aggressive. "O Jerusalem, Jerusalem, thou that killest the prophets, and stonest them which are sent unto thee."[1] Of course Jerusalem killeth her prophets. For what is a prophet? If he is a true prophet, is he not so because of his insight into life in general and into the inevitable consequences of our momentary passionate actions? Then because of this very insight he can never qualify as a popular leader, the hero of the passing moment. Popularity is hardly the rôle of a true prophet. Therefore Christ says:

> Woe unto you when all men shall speak well of you, for so did your fathers unto the false prophets.[2]

The greater a general insight is the more it is at variance with the vociferous passions of the moment. Now, when we come to the insight that Christ taught, it was so universal that it was not even understood by the moment. Only its points of variance were felt and resented by an aroused nation on the eve of its rebellion and its destruction. And Christ was crucified.

The kingdom was to be within us. The kingdom was a matter of attitude and of understanding. But the kingdom was also after all like a mustard seed, which is the smallest of seeds, but which grows in time.

> Another parable put he forth unto them, saying, The kingdom of heaven is like to a grain of mustard seed, which a man took, and sowed in his field: Which indeed is the least of all seeds: but when it is grown, it is the greatest among herbs, and becometh

[1] Matthew 23:37.
[2] Luke 6:25.

a tree, so that the birds of the air come and lodge in the branches thereof.[1]

And so after all is human assimilation of all knowledge, and all insight. It is a matter of slow growth.

[1] Matthew 13:31-32.

ROME'S FALL RECONSIDERED

THE great Roman writers with whom we are familiar seem to have been quite conscious of Rome's progressive disintegration. Testimony from the eyewitnesses of the process is, of course, of the utmost importance. Let us hear to what fundamental factors they themselves attributed the decline of their commonwealth.

Probably no handy quotation has pursued us through our school years with such persistence as Pliny's *"Latifundia perdidere Italiam, jam vero et provincias."* [1] The elder Pliny was not merely a man of great learning, but a much traveled statesman of large and varied experience. Is it not interesting that he does not present us with a catalog of factors that were leading Rome to its destruction? On the contrary, without any apology he is crisply pointing to one predominating factor, which he names. The large estates, the *latifundia,* were ruining Rome as well as its provinces.

More rhetorical in form, but similar in meaning, is the arraignment of the vast latifundia and their owners in Seneca's letters.[2] Seneca himself was one of the richest land-owners of Rome, but as a statesman he gave warning, in public, of what the wealthy land-

[1] Plinius, Historia Naturalis, xviii, 7.
[2] Seneca, Epistolæ, 89.

owners did not care to hear in private. Seneca asks: "How far will you extend the bounds of your possessions? An estate which formerly held a whole nation, is now too narrow for a single lord." [1] In fact, Cicero had already reported the statement of the tribune Philippus that the entire commonwealth could not muster two thousand property owners.[2] The concentration of landed property must have been impressive.

The latifundia, according to one view, therefore, were the cause of ruin; but there was a more popular version of the decline, namely, *corruptio*: the corruption of morals, the corruption brought by wealth, the corruption brought by poverty, the all-pervading moral corruption of Rome. Livy invites us to follow first the gradual sinking of the national character, later on the more rapid tempo of its downward course until the days are reached when "we can endure neither our vices nor their remedies." [3] And what great Roman of that period did not complain of corruption? Read Tiberius' famous letter to the Senate, which Tacitus has transmitted to us. The Senate complained of luxury and corruption and called on the emperor for action and Tiberius answered:

> That these excesses are censured at entertainments and in private circles, I know quite well. And yet, let a law be made with equal penalties, and the very men who call for a reform would be the first to make objections. The public peace, they would say, is disturbed; illustrious families are in danger of ruin. . . .[4]

[1] *Ibid.*
[2] "Non esse in civitate duo milia hominum qui rem haberent." Cicero, De Officiis, ii, 73.
[3] Livius, i, Praefatio.
[4] Tacitus, Annales, iii, 54.

Perhaps the most striking expression[1] of the progressive moral deterioration of the Romans is in Horace's ode "Ad Romanos":[2] Damnosa quid non imminuit dies? Aetas parentum pejor avis tulit nos nequiores, mox daturos progeniem vitiosiorem—"What does ruinous time not impair? The age of our parents, more degenerate than that of our grandfathers, produced us, even more worthless, and we shall give birth to a still more vicious progeny!" A cheerful prospect! But why such a note of despair? What is the cause of this moral corruption and degeneracy of which all Roman writers of the period complain?

In that very same ode Horace tells us why he takes so desperate a view of things. The great deeds of the Romans were the deeds of a sturdy farmer race: sed rusticorum mascula militium Proles, Sabellis docta ligonibus Versare glebas.[3] These farmers' sons existed no longer. If they could not maintain themselves on their farms, still worse were the chances for a respectable existence in Rome. There they lost what little they had and became demoralized, dependent paupers.[4]

The two complaints, the two Roman explanations of their own decline and disintegration *reduce themselves, therefore, to one single explanation.* For it is

[1] Among the picturesque characterizations of Roman degeneracy Columella deserves a very high place with his "Nam sic juvenum corpora fluxa et resoluta sunt, ut nihil mors mutatura videatur." "For so limp and dissolute are the bodies of the young men, that it does not seem as if death could make any change in them!" Columella, i, Praefatio.

[2] Horatius, Carmina, iii, 6.

[3] *Ibid.*

[4] Juvenalis, iii, 21 *et seq.;* Martialis, iv, 5.

clear that the latifundia and corruption are but different aspects of the same social phenomenon. If the moral disintegration was due to the disappearance of the self-supporting, self-respecting farmer class, and the inordinate wealth and fantastic luxury of the small upper class, *the latifundia were but the real estate expression of the same phenomenon.* Innumerable small farms had been replaced by extraordinarily large estates—the latifundia.

I do not doubt for a moment that the Romans were quite conscious of the connection between the latifundia and the corruption. Take, for instance, Sallust, who states it very clearly in his so-called epistles to Cæsar:

> When the people were gradually deprived of their lands, and idleness and want left them without a place to live on, they began to covet other men's property and to regard their liberty and the interests of their country as objects for sale. Thus the people who had been sovereign and who had governed all nations, became gradually degenerate; and instead of maintaining their common dominion brought upon themselves individual servitude.[1]

We are therefore justified, I believe, in stating that the contemporary witnesses of the decline of Rome had only one explanation of its cause; but while some emphasized its moral aspect and others its economic, still others, like Sallust or Pseudo-Sallust, have emphasized the political effect of the economic and moral disintegration of Rome.

The small farms disappeared. Why did they disappear? If we go back again to Roman literature to see just how the small farms disappeared and just how their place was taken by single latifundia, we find

[1] Sallust, Epist., i, 5.

little material that may be considered as a direct answer to our question. Such little material as we do find seems to suggest violence. Thus we are told in the Metamorphoses of Apuleius how the rich man, after despoiling his poor neighbor of his flocks, "resolved to dispossess him of his scanty acres and, inventing some empty quarrel over the boundaries, claimed the whole property for himself." [1] An intimation of similar proceedings is to be found in Sallust's "Jugurthine War":

> The parents and children of the soldiers, meantime, if they chanced to dwell near a powerful neighbor, were driven from their homes. Thus avarice, leagued with power, invaded, violated and laid waste everything without moderation or restraint, disregarding alike reason and religion till it rushed headlong, as it were, to its own destruction.[2]

Similar is the meaning of Horace's famous "Ode on Roman Luxury and Avarice":

> Quid, quod usque proximos
> Revellis agri terminos et ultra
> Limites clientum
> Salis avarus? Pellitur paternos
> In sinu ferens deos
> Et uxor et vir sordidosque natos.[3]

These and one or two similar stories [4] are about the only material at hand that bears directly on the wiping-out of small farms. Shall we therefore reach the conclusion that the innumerable small farms were wiped out by the violence of the few rich? Have we any other material relating to the life of the petty

[1] Apuleius, Metam., ix, 35.
[2] Sallust, Jugurt., xli.
[3] Horatius, Carmina, ii, 18.
[4] Quintilian, Apes Pauperis, xiii, 4; Seneca, Epist., 90, 39.

Roman farmer? None of any consequence. That being the case, it would be contrary to common sense to assume that small farming, once spread over the Italian countryside, was wiped out by violence. There is reproof and horror in the quotations cited; the authors are shocked, and we are shocked only by the unusual. Hence, the element of violence is emphasized in the material quoted, because it is exceptional. It takes too deep a mind to follow what is slow and uneventful, to find beauty in the deep ruts of a muddy country road, and treasures in the day-by-day life of the most humble. The story of the plain farmer we can expect to find in neither literature nor history. History as well as literature is a mountain-climbing expedition.

What do we know about Roman farmers that is not legendary in its nature? We know that in the earlier period of the Republic they considered seven jugera as ample for the support of a Roman farmer and his family. That is supposed to have been the size of the allotments after the expulsion of the kings; that was the size of the allotments in the colonies established by Manius Curius after his great conquests. It is he who is credited with the statement that "the man must be looked upon as a dangerous citizen, for whom seven jugera of land are not enough." [1]

Why did the seven-jugera farms disappear? Why was their place taken by the large private domains, the latifundia? That we are dealing here with the

[1] Plinius. Historia Naturalis, xviii, 4, 3. Also Valerius Maximus, iv, 3, 5. Sextus Aurelius Victor, De Viris Illustribus Urbis Romae, xxxiii, gives fourteen jugera as the size of the allotment.

fundamental problem of the Roman commonwealth is indicated by all its external struggles. Only glance at the way the problem is formulated by Tiberius Gracchus:

> The wild beasts of Italy have their dens and hiding places, while the brave men who spill their blood in her defense have nothing left but air and light. Houseless and without a spot of ground to rest upon, they wander about with their wives and children. Their commanders do but mock them when they exhort the soldiers in battle to defend their tombs and temples against the enemy; for out of so many Romans not one has a family altar or ancestral tomb. The private soldiers are called masters of the world, but fight and die to maintain the luxury and wealth of others; and they are called masters of the world without possessing a single clod to call their own.[1]

Unfortunately, about the process of the expropriation of the farmer class Tiberius Gracchus does not tell us any more than did Sallust or Pseudo-Sallust in his letter to Cæsar; but an interesting clue may be found in the size of the Gracchan allotments: they were to be thirty jugera each. Why not seven? Later on we know that the triumviral assignments were, according to Trentinus, fifty jugera, the assignments of Cæsar the dictator were sixty-six and one-third jugera. In the Augustinian colony, Emerita, we learn from Hyginus, the assignments were four hundred jugera.[2]

How then could seven jugera suffice for the farmers of early Rome? Did not the ancients speculate on the subject? Yes, they did. Pliny discusses this very problem. He is wondering about the productivity of the soil in the olden days, and here is what he tells us:

[1] Plutarch, Tiberius Gracchus, ix.

[2] Mommsen, Zum römischen Bodenrecht. Historische Schriften, v. 2, viii, 1, 81.

What, then, was the cause of such fertility? In those days the lands were tilled by the hands of generals, and there is reason to believe that the soil exulted beneath a ploughshare crowned with laurel, and a husbandman graced with triumphs; whether it is that they gave the seed the same care that they had given the conduct of wars, and arranged their fields with the same diligent attention as their camp, or whether it is that under the hands of honest men everything grows more gladly, since it is more carefully tended.[1]

Pliny may have been right in his explanation or he may have been wrong. The important thing is that the simple circumstance that a Roman could in former generations make a living on seven jugera distinctly required explanation.

There were other explanations. Is not one offered, for instance, by Lucretius' lines at the end of his second book? What does Lucretius tell us about mother earth in general and his Roman soil in particular?

At first she corn and wine, and oil did bear
And tender fruit, without the tiller's care;
She brought forth herbs, which now the feeble soil
Can scarce afford to all our pain and toil.
We labor, sweat, and yet by all this strife
Can scarce get corn and wine enough for life.
Our men, our oxen, groan, and never cease,
So fast our labors grow, our fruits decrease.
Nay oft the farmers with a sigh complain,
That they have labor'd all the year in vain,
And looking back on former ages, bless
With anxious thoughts their parents' happiness

. . . .

Content with what the willing soil did yield,
Though each man then enjoyed a narrower field.[2]

[1] Plinius, H. N., xviii, 4.
[2] Thomas Creech's translation. London, 1683, Book 2, l. 1111-1125.

This statement is of great importance, but only if corroborated by facts. To accept Lucretius' evidence by itself as sufficient and conclusive would be rather hasty. Lucretius as a philosopher is dealing here with the decay of the world, and hence the question naturally suggests itself: Did the actual situation, as he observed it, lead him to such a conclusion, or did his philosophy color his observations?

If the lines cited from Lucretius are a true statement of fact, the economic, and hence also social and political, effects of such conditions were bound to be so disastrous that it would be reasonable for us to expect a fairly general outcry.

A general outcry is, of course, important historical evidence, but it is not the cause that makes us complain, it is the effect, the situation in which we happen to find ourselves. A meat-market riot is as much concerned with the cause of the high cost of living, as the Roman

It is somewhat inaccurate, but it seems to me the most poetical translation of Lucretius.

Lucretius Carus, De Rerum Natura, Libri Sex, ii, 1157-1174.

Praeterea nitidas fruges vinetaque laeta
Sponte sua primum mortalibus ipsa creavit,
Ipse dedit dulcis fetus et pabula laeta;
Quae nunc vix nostro grandescunt aucta labore,
Conterimusque boves et viris agricolarum,
Conficimus ferrum, vix arvis suppeditati:
Usque adeo parcunt fetus augentque labore.
Iamque caput quassans grandis suspirat arator
Crebrius, incassum manuum cecidisse labores,
Et cum tempora temporibus praesentia confert
Praeteritis, laudat fortunas saepe parentis.
Tristis item vetulae vitis sator atque victae
Temporis incusat momen, caelumque fatigat,
Et crepat, antiquum genus ut pietate repletum
Perfacile angustis tolerarit finibus aevom,
Cum minor esset agri multo modus ante viritim:
Nec tenet omnia paulatim tabescere et ire
Ad capulum, spatio aetatis defessa vetusto.

sumptuary and moral laws were concerned with the cause of the much-complained-of corruption. The effects one can always see; of the effects one constantly hears: but the cause one must find. So it is, so it was; hence Virgil's "Felix qui potuit rerum cognoscere causas."

Nor are we always primarily interested in the true cause. There are situations where one is inclined to search for a life-preserver rather than for the cause of shipwreck. It is therefore wise to remember that the attitude of the impartial onlooker is likely to be quite different from that of the dramatis persona. Goethe indicates this difference:

> Wherein do gods
> Differ from mortals?
> In that the former
> See endless billows
> Heaving before them;
> *Us* doth the billow
> Lift up and swallow
> So that we perish.

So it happens that the true causes of things are hardly discussed in the markets and meeting-places. It is the future, not the past, that worries politicians. Remedies, not causes, are what they are bound to discuss. For life is purposeful, and only to its dissector is it a chain of causes. But Rome was not without dissecting scholars.

Let us therefore go and see how the great agricultural scholars of the time analyzed the situation. Let us read thoughtfully the writings of Columella. He was writing under the Principate, about 60 A. D. How does he begin his work? The preface begins:

I frequently hear the most illustrious men of our country complaining that either the sterility of our soil or the unseasonable weather has for many years been diminishing the productivity of the land. Others give a rational background to their complaints, claiming that the soil became tired and exhausted from excessive productivity in the past, and hence can no longer furnish sustenance to mortals with its former liberality.[1]

Columella does not agree with such a point of view. He ascribes the lack of productivity to poor farming; and hence he gives us instructions how to farm well. But is it not of the utmost significance that he published a voluminous treatise distinctly directed against a prevailing exhaustion-of-the-soil theory? These opening lines of Columella are far from accidental. Furthermore we learn from him a very important thing, and that is, *that nearly all agricultural writers of antiquity* (whose writings are lost to us) *viewed their contemporary agricultural situation as due to the exhaustion of the soil;* or, as they put it, as the result of the soil's old age.

Attention is called to the opening paragraph of Columella's second book, chapter i:

You ask me, Publius Silvinus—and I hasten to reply to you—why I began my former book *by practically contradicting the early opinion of nearly all writers on agriculture, and by rejecting as false their idea that the soil, worn out by long cultivation and exhausted, is suffering from old age.*[2]

[1] Columella, De Re Rustica, Lib. i, Ad Pub. Silvinum, praefatio: "Saepe numero civitatis nostrae principes audio culpantes modo agrorum infoecunditatem, modo caeli per multa jam tempora noxiam frugibus intemperiem: quosdam etiam praedictas querimonias velut ratione certa mitigantes, quod existiment, ubertate nimia prioris aevi defatigatum et effoetum solum nequire pristina benignitate praebere mortalibus alimenta."

[2] Columella: ii, 1: "Quaeris ex me, Publi Silvine, quod ego sine cunctatione non recuso docere, cur priore libro veterum opinionem fere

Hence we learn that the ideas of Lucretius were not peculiar to him alone, but (if we accept the testimony of Columella) they were the common conception of nearly all who seriously thought about and scientifically discussed the agricultural affairs of antiquity. If the works to which Columella is referring had survived and had been preserved to us, there would have been little left for us to discuss.

Columella refuted the exhaustion-of-soil conception. Let us see how he did it. We find in his book three arguments. First of all, the Creator has bestowed upon soil perpetual fecundity; hence it is impious to regard the soil as affected with sterility as with a disease. Divine and everlasting youth was allotted to our common parent, mother-earth; it is silly to assume that she is ageing like a human being.[1]

His second argument is particularly directed against Tremellius, whose writings (lost to us) he especially esteems. Tremellius is of the opinion that mother-earth has, like a woman, reached that point of her life when sterility takes the place of her former fecundity. To this Columella replies that he would have accepted Tremellius' view had the soil been completely unproductive. But he argues that we do not regard a woman as having reached the barren age, simply because she no longer gives birth to triplets and twins. Furthermore, when a woman has reached that age, the bearing of children cannot be restored

omnium, qui de cultu agrorum locuti sunt, a principio confestim repulerim, falsamque sententiam repudiaverim censentium, longo aevi situ longique jam temporis exercitatione fatigatam et effoetam humum consenuisse."

[1] Columella, i, praefatio.

to her, while the land, if abandoned for a time, will be found upon the return of the cultivator more fertile.[1]

And, finally, Columella is therefore of the opinion that the soil would never diminish in its productivity if properly taken care of and frequently manured.[2]

Very sound and sensible was his conclusion, but whether his advice could be followed is another question. What interests us is to ascertain not the state of the theory, but the state of the actual practice. That is the crucial and deciding question. Fortunately Columella answers this question in the third chapter of his third book, where he urges going into vine culture rather than into the cultivation of grain, because in the greater part of Italy no one can recall when grain produced four-fold. "Nam frumenta majore quidem parte Italiae quando cum quarto responderint, vix meminisse possumus."[3] In other words, if it did not produce four-fold, it produced three-fold or two-fold. In what situation would our modern farmers be, if the average productivity of their wheat, corn, barley, etc., should be somewhere between four and six bushels an acre, a productivity which would completely assure and enforce the abandonment of farming? But the Romans were in a worse situation. We now plough in a nine- or ten-hour day about two acres with an average team, even in heavy clay or sod. The Romans ploughed in light soil a jugerum, which is five-eighths of an acre; in heavy soil but half a jugerum. But that is not all; we plough but once;

[1] Columella, ii, 1.
[2] *Ibid.*
[3] Columella, iii, 3.

they, for lack of effective harrows, had to plough corn land anywhere from five to nine times. Now one can figure out where the Italian farmer found himself in the days of Columella! The game was up; but what stopped it? Many are the answers. Some tell us it was stopped by constant warfare. But Columella complains that the old Sabine quirites and Roman ancestors, in spite of the fire and sword to which they themselves were subjected, and in spite of the hostile invasion which laid waste their fields, nevertheless laid up greater store of corn than his contemporaries were able to do, although during the long-continued peace they might have improved their agriculture.[1]

When Columella wrote in A. D. 60, Italy certainly was enjoying a long protracted peace. Furthermore, one must remember that war as such, even if it should drive the farmers away from the land and keep them from cultivation for years, does not in any way exhaust the soil. For if the soil is not exhausted it will grow over with weeds and bushes, which will prevent the washing-away of the top soil, and when again put under the plough, the farmer will find his soil improved, because of the decayed weeds and other vegetable matter. If, on the other hand, the soil was abandoned when substantially so exhausted that it would not readily cover with weeds, then the top soil would gradually wash off and make its reclaiming difficult and costly.

[1] "Veteres illi Sabini Quirites atavique Romani, quamquam inter ferrum et ignes hosticisque incursionibus vastatas fruges, largius tamen condidere quam nos, quibus diuturna permittente pace prolatare licuit rem rusticam." Columella, i, praefatio.

The writers of the Principate look back to the sturdy past of the days of Cato the Censor. They were mistaken. For even in Cato's day agriculture had already declined in the greater part of Italy. His *Husbandry,* the earliest Roman agricultural book that has come down to us, practically disregards the cultivation of grain crops. His attention is devoted to the cultivation of the vine and olive.

Cato, when asked what is the most profitable thing in the management of one's estate, answered: "Good pasturage." "What is the next best?" "Fairly good pasturage." "What is the third best?" "Bad pasturage." "What is the fourth best?" "Tilling the soil."[1]

Such a statement requires no comment. And as a matter of fact, even in Cato's day Italy had to rely upon Sicily as its granary.[2]

Again it has been said that the free distribution and sale of corn at low prices in Rome ruined Roman agriculture. Mommsen takes this attitude in his Roman history, and it is generally accepted—but he is taking the effect for the cause. Importations of corn began when relief had to be given to the growing proletariat of Rome. Mommsen himself, in a later piece of research, admits that one often hears of high prices, and only exceptionally of low prices, of corn so that

[1] Cicero, De Officiis, ii, 25: ". . . illud est Catonis senis: a quo cum quaereretur quid maxime in re familiari expediret, respondit: 'Bene pascere'; quid secundum: 'Satis bene pascere'; quid tertium: 'Male pascere'; quid quartum: 'Arare.' "

[2] Cato called Sicily the nourisher of the Roman people "nutricem plebis Romanae Siciliam nominabat." Cicero, In Verr., ii, 2. Livy called even the single town of Syracuse: "Horreum atque aerarium quondam populi Romani." Livius, xxvi, 32.

as a whole rather too little than too much was produced.[1]

No evidence has come down to us that would indicate difficulties in disposing of grain; all the bitter complaints that we hear are about hardships and difficulties in *raising* grain. Look at Cato. What he thought of tillage we have heard. Yet Plutarch tells us: "As Cato grew more eager to make money he declared that farming was more an amusement than a source of income and preferred investing his money in remunerative undertakings, such as marshes that required draining. . . ."[2]

Here is the story in a nut-shell. An undrained marsh has never been tilled, and therefore never robbed of its fertility. Since one would hardly select low flats for vineyards, which require at least a slope, it is obvious that Cato drained the marshes for purposes of tillage. The initial expenses of drainage are heavy, yet Cato regarded the results as very remunerative, and that, in spite of Sicilian corn on the Roman market. To drain a rich marsh was obviously easier for the Romans than to reclaim large tracts of ordinary exhausted soil.

It is interesting that the lands that were first taken up by Roman cultivators were also, judging from our sources, the first to be exhausted. It was in Latium, where once seven jugera were ample to support a

[1] Mommsen, "Boden- und Geldwirtschaft der römischen Kaiserzeit" in his "Historische Schriften," Bd. ii, xxxvii, p. 604; "Es ist nicht selten von teuren Kornpreisen, nur ausnahmsweise von besonders niedrigen die Rede, so dass im Ganzen wohl eher zu wenig als zu viel produciert ward."

[2] Plutarch, Marcus Cato.. xxi.

family, that Varro finds an example of notoriously sterile soil. He mentions Pupinia in Latium: "Witness Pupinia, where the foliage is meager, the vines looked starved, where the scant straw never stools, nor the fig tree blooms, and trees and parched meadows are largely covered with moss." [1]

Two hundred years later Columella no longer singles out Pupinia, but refers to entire Latium as a country where the population would have died of starvation, had it not been for imported grain. So it came to pass

> that in that very land, in Saturn's own country, where gods taught their children how to till the soil, there at public auction we have to contract for corn imported from provinces beyond the seas, that we may not suffer from starvation, and wine we have to import from the Cyclades, from the regions of Boetia and Gaul.[2]

As the productivity of the soil diminished, and the crops could no longer repay the laborer, then the same process that occurred in England in the 15th and 16th centuries, the turning of arable land into pasturage, began in Italy, about two centuries before Christ. In Rome, too, this process was met by hostile legislation, as was the case in England, but without avail. As in England, so in Rome, it became a matter not of choice but of necessity, although even the thinking heads of both nations refused to admit it at the time,

[1] "Ut in Pupinia neque arbores prolixas, neque vites feraces, neque stramenta videre crassa possis, neque ficum mariscam, et arbores plerasque, ac prata retorrida, et muscosa." Varro, De Re Rustica, i, 9.

[2] "Itaque in hoc Latio et Saturnia terra, ubi dii cultus agrorum progeniem suam docuerant, ibi nunc ad hastam locamus, ut nobis ex transmarinis provinciis advehatur frumentum, ne fame laboremus; et vindemias condimus ex insulis Cycladibus ac regionibus Boeticis Gallicisque." Columella, i, praefatio.

and preferred to ascribe the change to greed and corruption. In England they blamed the poor sheep; in Rome they blamed the attractions of city life. So we hear Varro lamenting:

> Our very corn that is to feed us has to be imported for us from Africa and Sardinia, while our vintages come in ships from the islands of Cos and Chios. And so it happened that those lands which the shepherds who founded the city taught their children to cultivate are now by their descendants converted out of greed from cornfields back into pastures, violating even the law, since they fail to distinguish between agriculture and pasturage, for a shepherd is one thing and a ploughman another.[1]

It seems to me that the progressive exhaustion of Roman soil is, judging by all the sources at our disposal, completely established; but there prevails in literature a diametrically contrary version of the story—that of Rodbertus, who is regarded by economists as their authority. Rodbertus, too, quotes Columella's statement about crops not producing in Italy the fourth grain. He also refers to Varro's statement quoted above, but he explains it all "propter avaritiam." It is through avarice that all good soil was put under pasturage, because cattle-raising paid better. The fact that soil produced next to nothing when cultivated is explained by him thus: only the very poorest soil was under the plow, because wine, oil, and fruits were so much more profitable. The type of production

[1] ". . . frumentum locamus, qui nobis advehat, qui saturi fiamus ex Africa, et Sardinia; et navibus vindemiam condimus ex insula Coa, et Chia. Itaque in qua terra culturam agri docuerunt pastores progeniem suam, qui condiderunt urbem, ibi contra progenies eorum, propter avaritiam contra leges ex segetibus fecit prata, ignorantes non idem esse agriculturam et pastionem. Alius enim opilio, et arator." Varro, ii, De Re Pecuaria, praefatio.

changed, but became by no means worse, and agriculture was certainly not to blame, if Italy was not producing its own grain. "Der Ackerbau selbst war also unschuldig daran, wenn Italien nicht mehr seinen ganzen Getreidebedarf lieferte." [1]

The statement of Rodbertus' can with difficulty be taken seriously. First of all, the Romans not only failed to produce their grain; they failed to produce their vintage as well, in spite of the premium put on Italian wine by prohibiting the planting of vines in Gaul. Secondly, as a farmer, Rodbertus must have realized that if they practised rotation of crops, which he assumes, the fact that the Romans of Columella's time could not produce a fourth grain would indicate sterility, not of their poorest field, but of all their arable fields. Thirdly, to assume that the Romans would select their very worst fields, not out of necessity but out of choice, that they would be satisfied to plough and work and harvest those fields for a gain of one or two bushels over and above the bushel of seed, is to assume that the Romans had become insane.

The soil of Italy did not get exhausted over night. It was a long process and many were its stages. Besides, exhaustion is a very relative term; not only relative from an agro-technical point of view, but also relative to the physical needs as well as the economic capacities of the owner.

The expropriation of the Roman peasantry, the concentration of ownership of land in the hands of the few, to which the Romans ascribed the ruin of the

[1] Rodbertus, Zur Geschichte der agrarischen Entwickelung Roms. Hildebrand's Jahrbücher Bd. ii, 1864, pp. 218-19.

Empire, is also a very gradual process and runs parallel with the process of soil exhaustion. Compared with the seven-jugera holdings of the early Republic, the hundred- or hundred and fifty-acre plantations to which Cato refers are large estates. These "estates" of Cato, which in size correspond to an average American farm, gradually disappear and their place is taken toward the end of the Republic and under the Principate by vast domains measuring thousands and thousands of acres. The process of transformation was slow but constant. This process was not only agonizing to the people; it was sapping the very life of Rome as a nation, decreasing its population, undermining its morale and convulsing its political fabric. The beginnings of this process are almost lost in the darkness of Rome's legendary period. For the first violent expression of Roman social life to which we are introduced is the outcry of the indebted and bonded farmer class. In the growing wholesale indebtedness of the Roman farmer some historians have seen the key to Rome's political struggles, but the cause of this indebtedness was either not discerned or was viewed more or less as a mystery. So, for instance, Büchsenschütz tells us that if the origin and character of the debts are veiled for us in darkness, the fact remains that the plebeians were the debtors and the patricians the creditors. The division between rich and poor coincides with the division of the orders, and the struggle of the debtors against the creditors was therefore fought out as a purely political struggle.[1]

[1] "Ist somit die Enstehung und das Wesen dieser Schulden für uns sehr im Dunkel gehüllt, so wird die Sache noch dadurch bedenk-

There are two kinds of indebtedness: debts for productive purposes and debts for purposes of consumption. If the reports of American banks should show a growing extension of credit it would be fair to assume a growing extension of industry, because banks as a rule deal with credit for productive purposes only. If, on the other hand, the report should reach us that the volume of business of our pawnshops had greatly increased, it would indicate that poverty is on the increase, that the incomes of the borrowers are insufficient to meet their ordinary expenses and hence that they are borrowing for purposes of consumption. The wholesale indebtedness of the Roman farmer class obviously suggests indebtedness for purposes of consumption.

If the farmer is borrowing to meet the exigencies of a so-called bad year, his distress is temporary, and he is likely to square himself during the next good year; but if his distress is due to the progressive deterioration of his farm, he will be unable to extricate himself. Such indebtedness is hopeless. The increasing weight of accumulated interest on the loan and the decreasing productivity of the land seal the fate of the landowner. He certainly is not in an economic position to increase his land-holdings to a point where the larger product might supply his wants. Because

licher, dass als Schuldner die Plebejer, als Gläubiger die Patricier erscheinen, der Unterschied von reich und arm genau mit dem Unterschiede der Stände zusammenfällt und infolge dessen der aus der Schuldnot entstandene Zwist zu einem Kampfe der Stände gestaltet und lediglich als solcher ausgefochten wird." B. Büchsenschütz, Bemerkungen uber die römische Volkswirtschaft der Königszeit (Wissenschaftliche Beilage zum Program des Friedrichs-Werderschen Gymnasiums zu Berlin 1886), p. 34.

he does not have enough land, what little he has will be taken from him and be given to him that has both land and economic capacity. In this way a farmer will be driven off the land and the holdings of some one else increased. That is the process of concentration of landed property. If this process should appear as a general phenomenon, as it did in both Rome and Greece, it would be a factor of momentous social significance.

The entire history of Rome is but a series of illustrations of this story. Steady is the legislation against interest and drastic are the measures against the money lenders, but even in spite of social revolutions and social wars, the concentration of landed property is unchecked. Because of this peculiar character of credit in certain historical periods, money-lending was not a savory occupation. The gentleman, therefore, who in our industrial and mercantile life is a pillar of society and a respectable financier, is known by a different name in agricultural communities. His name is *Usurer*. Not that his profits from money-lending are any larger, but that he is lending money for purposes of consumption to a man as a rule already economically doomed, while the "banker" is lending money for productive purposes and as a rule to the advantage of the borrower. Hence the different attitude towards the "financier" now and in ages past. Cato, when asked what he thought of money-lending, answered: "What do you think of murder?" [1]

One must not, however, get the idea that all con-

[1] Cicero, De Officiis, ii, 25.

centration of real property is necessarily due to indebtedness. Just as one is as a rule unwilling to part with a lucrative piece of property, one may be willing and anxious to part with property that is unproductive, and that even without pressure of debts.

We have pretty illustrations of this in Roman literature. Cicero, in his second harangue against P. S. Rullus, tells us that Publius Lentulus was sent by the Senate to purchase a private farm in Campania that projected into the public domain, but he was unable to purchase the farm for any money, because its owner could not be induced to part with his most productive parcel of land.[1] Let us read carefully, on the other hand, a most remarkable letter of the younger Pliny to Calvisius Rufus, from whom he is soliciting advice. The younger Pliny is contemplating the increase of his latifundia by adding an enormous neighboring estate, which is offered as a bargain. Here are two extracts from his letter:

> I feel tempted to purchase, first, because of the convenience and pleasure of uniting adjacent estates. . . . But the fertility of the land is overtaxed by poor tenants. For the last proprietor was constantly selling their whole stock, and though he reduced the arrears of the tenants for the time, he weakened their efficiency for the future, and as their capital failed them their arrears once more began to mount up. I must therefore set them up again; and it will cost the more because I must provide them with honest slaves; slaves working in chains, I do not own any, nor does any landowner in that part of the country. Now let me tell you the price at which I think I can purchase the property. It is three million sesterces, though at one time the price was five, but owing

[1] Cicero, De Officis, ii, 25.

to the poverty of the tenants and the general difficulty of the time the rents have fallen off and the price has therefore dropped also.[1]

Here we have the typical latifundia of imperial Rome, sublet to tenants. They could not pay their rent; the owner thereupon sold their stock, which did not strengthen their productive and paying capacity. The fertility of the estate is admittedly impaired by this lack of stock and it is offered as a bargain. Is the younger Pliny attracted by it as a money-making proposition? Hardly. But "prædia agris meis vicina atque etiam inserta" and the "pulchritudo iugendi"—the old story of rounding up one's estate, by buying the adjoining one. The elder Pliny in telling us how the latifundia were ruining Rome must have had in mind precisely such purchases as his nephew contemplated, for without any too obvious connection he adds: "With that greatness of mind which was so peculiarly his own, Cnaeus Pompeius would never purchase the land that belonged to a neighbor." [2]

Still, do not let us simplify the process of concentration too much. It undoubtedly had as an underlying cause the relative unproductivity of the soil. The process of concentration followed many parallel routes. Indebtedness was undoubtedly the greatest factor in abolishing small holdings. Unproductivity of agriculture naturally led to cattle-ranches, which required much larger holdings. Wealthy men acquired and accumulated vast domains rather for the pleasure of possession than as a paying investment. But the process of deterioration went on, and legislative inter-

[1] Plinius Secundus, Epist., iii, 19.
[2] Plinius, H. N., xviii, 7.

ferences could neither stop the robbing of the soil nor the depreciation of land values. Negligent cultivation of one's own land was punishable, so was conversion of arable land into pasturage; but neither law proved effective. To maintain land values, as early as 218 B. C., the Claudian law excluded senatorial houses from mercantile occupations and compelled them to invest in Italian land. After Trajan's time, one third of their wealth had to be invested in land. Tiberius, in A. D. 33, put in force an old law and compelled all bankers to invest, so far as can be made out, two-thirds of their working capital in Italian lands.[1] Such measures maintained for a time the land values, but they could not touch the underlying cause—the process of spoliation and exhaustion of the fields as well as the process of proletarization, corruption and depopulation of the nation.

Some questions suggest themselves in this connection. First of all, did not the Romans know how to conserve or improve their soil and thereby make their agricultural labor more productive? The answer to this can be only that the Roman knowledge of rational and intensive agriculture was so great as to be fairly startling. The knowledge of the Roman Scriptores Rei Rusticae is superior to any agricultural practice of the Middle Ages or even of modern Europe at the beginning of the 19th century.

Why then did the Roman farmers fail to improve their methods of agriculture, even when pressed by necessity to do so, even when threatened with extermi-

[1] See Mommsen, Boden- und Geldwirtschaft der römischen Kaiserzeit. Historische Schriften, Bd. ii, xxxvii, p. 595.

nation? It was easier said than done. Behind our abstract agricultural reflections are concrete individual farms. The owners of the rundown farms are impoverished, and when a farmer is economically sinking he is not in a position to improve his land.

Only one with sufficient resources can improve his land. By improving land we add to our capital, while by robbing land we add immediately to our income; in doing so, however, we diminish out of all proportion our capital as farmers, the productive value of our farm land. The individual farmer can therefore improve his land only when in an economically strong position. A farmer who is failing to make a living on his farm is more likely to exploit his farm to the utmost; and when there is no room for further exploitation, he is likely to meet the deficit by borrowing, thus pledging the future productivity of his farm. Such is the process that as a rule leads to his losing possession of his homestead and his fields, and to his complete proletarization.

The exceptional man might pull himself up under adverse conditions, but, on the other hand, a man of such exceptional resourcefulness and ability will not permit such deterioration of his farm. But whatever the exceptional man may or may not do, here we are dealing with the average men, the habitual victims of circumstances. In this connection let me point out that already Columella as well as Tremellius fully realized the situation, and they therefore regarded the ability to lay out money as the essential condition for improved agriculture. "For neither knowledge nor effort can be of any use to any person whatsoever, without those expenses which the operations require." [1]

The great agricultural knowledge of the Romans must not, however, be dismissed lightly. It opens up many serious questions. All that is implied by the agricultural revolution, the seeding of grasses and legumes, the rotation of crops, yes, even green manuring, all that was perfectly known to the Romans; why was it not practised for two thousand years or more? I do not know. It shows up so-called intellectual knowledge in rather an unenviable light, but that is not an answer.

The only interesting and important conclusions we may draw from the agricultural history of antiquity, the Middle Ages and modern times is that the talk about agricultural evolution from an extensive .to an intensive culture belongs in the class of generalizations which should not be taken seriously. The opposite development is the more likely—probably the development from the intensive garden plot culture to extensive agriculture. Thus the farming of the Romans on seven-jugera farms was, like the farming of the Chinese and Japanese, very intensive, their small grain fields being planted in rows, hoed and weeded and carefully manured with excrements and ashes and stable dung. The experience of China and Japan has proven that on very small land plots such intensive agriculture can maintain itself indefinitely without any

[1] "Qui studium agricolationi dederit, antiquissima sciat haec sibi advocanda, prudentiam rei, facultatem impendendi, voluntatem agendi. Nam is demum cultissimum rus habebit, ut ait Tremellius, qui et colere sciet et poterit et volet. Neque enim scire aut velle cuiquam satis fuerit sine sumptibus, quos exigunt opera." Columella, i, 1.

recourse to scientific repletion of the soil by mineral fertilizers. Why did Rome fail, where China and Japan succeeded, after a fashion? We do not know. Certain it is that the intensive agriculture in Rome was ill-fated and that Virgil was well justified in drawing from it the conclusion: "Thus fate drags all to ruin with a backward pull, as when a rower hardly drives his boat against the stream: if once he drop his arms, forthwith the rushing current whirls him down."

Already in Cato's time the growing of grain crops was so utterly unprofitable that he did not even take the trouble to instruct us on this point. All he could tell us was that plowing was less profitable than the worst pasture. In the case of Varro, instruction on intensive grain-raising is a tradition of earlier times. His preface to the second book frankly admits that there is no use talking of crop-raising, when agriculture has been abandoned for grazing. In the case of Columella the literary tradition is still more pronounced. The estates, he tells us, are as large as provinces; nowhere in Italy in the memory of mankind have they raised a four-fold grain crop; yet he outlines for us a most intensive culture of grain which is evidently a long extinct tradition. The mistakes that he occasionally makes also prove that he never was an eyewitness to the operations, in spite of his wide experience. Thus he suggests seeding alfalfa one cyathus for fifty square feet, which amounts to several bushels per acre—an impossible proposition. The well-known botanist, Mattioli, who wrote in the 16th century, tells us that while alfalfa was obviously grown once upon a time in Italy, he has never found any one who has seen the plant in its seed.[1] I believe that Columella was even in his own time in Mattioli's position.

In her early days Italy was famous for her wheat, which provided not only her own population but also that of Greece. The fertility of Italian soil was probably the reason for the establishment of Greek colonies in southern Italy. The importation of Italian wheat into Greece in Sophocles' time is still famous. But in Cato's time Italy was already dependent upon Sicily, which Rome's great old men called the provider for the Roman people. In all probability this dependence upon Sicily as its granary was the paramount reason for Rome's conflict with Carthage. Province after province was turned by Rome into a desert, for Rome's exactions naturally compelled greater exploitation of the conquered soil and its more rapid exhaustion. Province after province was conquered by Rome to feed the growing proletariat with its corn and to enrich the prosperous with the loot. The devastations of war abroad and at home helped the process along. The only exception to the rule of spoliation and exhaustion was Egypt, because of the overflow of the Nile. For this reason Egypt played a unique rôle in the Empire. Tacitus tells the story in a nut-shell.

[1] Pietro Andrea Mattioli, Discorsi ne i sei libri di Dioscoride. Venetia (1621) iii, 117, p. 491. Ordinary varieties of clover were regarded by the same writer as rare medicinal plants. So Trifolium Asphaltine could occasionally be found on uncultivated fields near Lucca, Trifolium pratense near Naples. See also the German edition of Mattioli's "Kreuterbuch." (1586) iii, 86, p. 291.

He is describing Germanicus' travels in the east, including Egypt:

Another point appeared to him [Tiberius] of greater moment. Among other rules, Augustus had established the maxim of state policy that Egypt should be considered forbidden ground, which neither the senators nor the Roman knights should presume to tread upon, without the express permission of the prince. This was no doubt a wise precaution. It was seen that whoever made himself master of Alexandria, with the strongholds which by sea and land were the keys of the whole province, might with a small force, make head against the power of Rome, and, by blocking up the plentiful corn country, *reduce all Italy to a famine.*[1]

To provide with grain the dwindling population of Italy was a life and death question to the Empire even in the days of Tiberius, and Tiberius freely admitted it. When his attention was called by the ædiles to the growing luxury of the rich and their breaking the sumptuary laws Tiberius answers:

But after all, is the mischief, of which the ædiles complain, the worst of our grievances? Compare it with other evils, and it vanishes into nothing. But no one considers how much Italy stands in need of foreign supplies, and how the commonwealth is every day at the mercy of the winds and waves? The produce of colonies is imported to feed the landlord and his slaves. Should these resources fail, will our groves and our villas support us? That care is left to the ruler. Should he neglect it, the commonwealth would be lost.[2]

The commonwealth was not yet lost in Tiberius' days, but it was already doomed and Rome knew it. The fundamental trouble could not be cured. In Italy,

[1] Tacitus, Ann., ii. 59.
[2] Tacitus, Ann., iii, 54.

labor could not support life, and men and women were not reared to maintain the population and Rome's dominion over the world. Yet Rome's livelihood was its dominion; it lived, as Seneca put it, on "the spoils of all nations." And in view of the fact only too obvious that Italy's population was dwindling, it was quite natural for Seneca to point out to his fellow-citizens, that what one nation has taken from all nations, may be retaken even more easily by all of them from the one.[1]

But we are told that Italy's depopulation was due to the civil strife and wars, to the ever-increasing marsh areas, and growing unhealthiness, and to a thousand and one other cherished explanations—all of them to a large extent based on contemporary documents and to a greater or lesser extent true, but all of them, at best, important symptoms or minor effects rather than fundamental causes. Thus they ascribe the lack of daily bread from which Rome was suffering to the circumstance that Italy found it more profitable to grow vines, to go into grazing and leave the grain production to its provinces. Of course, if in Nero's days the fourth grain could not be produced anywhere in Italy, it is perfectly true that it was profitable for Italy to leave grain production to the provinces. To claim for Italy a choice in the matter is somewhat misleading.

Misleading as well is the talk about economic differentiation: Italy producing this or that, while Africa or Spain or Sicily produced grain. The truth is that

[1] Seneca, Epist., 87.

the granaries of Rome, with the exception of Egypt, were undergoing the same process of exhaustion and devastation. Recall Sicily, Sardinia, northwestern Africa and Spain, not to mention Greece which antedated even Italy in her exhaustion. Neither can the opinion be taken under serious consideration which regards the growing insalubrity of Italian lowlands as the cause of depopulation, which led to undermined national strength, a diminishing agricultural area, etc. For those who hold such an opinion seem to forget that many other provinces of the Empire underwent the same process of rapid depopulation without turning into swamps, but rather that many parts of them, like Libya, for instance, were turning into arid deserts. As a matter of fact, the same fundamental causes that were increasing the swampiness of Latium and Campania were turning northwestern Africa and portions of Asia into deserts. These causes were, or this cause was: *agri deserti*—abandoned fields. Any farmer who is working his fields takes care that they shall drain properly. On any large farm one has dry land and wet land. Some lands drain rapidly and are fit for early plowing and sowing, while other lands perhaps situated in hollows remain wet for a longer time. There are many ways in which the farmer attends to a more or less proper drainage. Only advanced farmers in modern times are putting in tile drains that do not choke and that permanently improve wet or springy land. As a rule the farmer in times past did what most farmers do now. They drained wet fields every year by draining furrows perhaps a week or so before plowing time, or if the situation required they drained them with somewhat more permanent open ditches, or they drained the superfluous water with deep ditches filled with stones and covered with earth. In England and in western Europe the medieval farmer was cultivating most of his land in ridges—an extremely wasteful system of drainage. You can find pictures of it in medieval manuscripts. Just as every road, as a matter of routine, is so built as to allow the water to drain, so is minor drainage a matter of daily farming routine to such an extent that the farmer is actually not even conscious of it. But all this minor everyday drainage has to be repeated every year; even more permanent drainage, such as the open ditches, must be kept open, and stone ditches have to be watched lest they fill up with dirt and choke. But what happens when the fields fail to reward labor and are abandoned? If the highlands are not capable of covering themselves readily with vegetation, the top soil is washed away and a desert is left, while the deserted lowlands with clogged-up drainage are bound to turn swampy and unhealthful. That swampy lands breed mosquitoes and that mosquitoes are responsible for malaria the Romans seem to have known quite well. Already Varro had advised against the use of slaves for work in marshy lands, and advocated taking chances with the health of hired freemen;[1] for, as he explains in a previous chapter, "animalia quaedam minuta" that breed in swamps cause grave maladies.[2]

[1] Varro, i, 17.
[2] *Idem*, i, 12.

Columella describes the mosquito in still more unmistakable terms: "infestis aculeis armata animalia." [1]

It is therefore only reasonable to assume, so far as Campania and Latium are concerned, that the population was not driven out by malarial mosquitoes, but that mosquitoes took peaceable possession of the lands already abandoned by their cultivators. Thus in the year 395 the abandoned fields of Campania alone amounted to something over 528,000 jugera.[2]

Much more plausible is the opinion that the depopulation of Italy as well as the devastation of its fields was due to wars and civil strife. This argument is self-evident and can be supported by endless quotations from the great Roman authors. Obvious as the argument is, it does not, however, stand up well under further scrutiny. For the analysis of the actual data at our command is most perplexing. The steady shrinkage of population in the ancient world did not follow, curiously enough, in the wake of its bloodiest wars, but in times of complete peace. The fearful losses of Rome's greatest wars, on the other hand, losses, for instance, occasioned by the Punic Wars, were rapidly made up, and in spite of further wars the population was steadily increasing. The same was true about the temporary decrease of population occasioned by a plague. Different was the situation in the period under discussion. Losses occasioned by wars and plagues were never made up, and during the longest and profoundest peace that Rome ever enjoyed the Roman population was steadily shrinking and its na-

[1] Columella, i, 5.
[2] Cod. Theodos., xi, 28, 2.

tional strength steadily melting away. Quite similar was the process of depopulation in Greece. There, too, the losses of the bloodiest wars were made up and the country was becoming depopulated under most peaceful auspices. These, I believe, are facts generally acknowledged by historians holding diverse points of view. Professor Eduard Meyer sums up the situation as follows: [1]

> There is no doubt that the two hundred years of peace brought about by the Principate at first effected an increase of population in all parts of the Empire, except where, as in Italy, Sicily, Greece, the economic and social conditions interfered or even enforced a backward tendency. In northern Africa, Spain and Gaul, in the Danube provinces, in Asia Minor, in Syria, we therefore find signs of prosperity everywhere. All the more striking is the inner dissolution of antique life and culture which followed so soon afterward. Thus, after all, during two centuries of peace, under a careful and circumspect government, the wealth and the population and even to a much greater extent, the very capacity of the Empire, were dwindling continuously. The economic decline that had devastated Greece and Italy spread thence to the provinces of Sicily, Spain, Africa, Gaul, one after another. The great civilized state [*Culturstaat*] was hardly capable any longer of raising armies to hold down barbarian tribes like the Marcomanni. The devastation wrought by the plague in the population of the Empire under the Emperor Marcus was never overcome. Our sources, it is true, will not permit us to give this development statistical expression. The terrible struggles of the third century, the continual uprising of the armies and provinces against each other hastened the process and completed the downfall of the State. The new State erected by Diocletian and Constantine saved, it is true, the ruins of ancient civilization and gave new stiffening to the East. It could not, however, maintain the West against

[1] Ed. Meyer. "Die Bevölkerung des Altertums," Handwörterbuch der Staatswissenschaften, 3te Auflage (1909) Bd. ii, pp. 911-912.

the barbarians, who had first been called as soldiers to the empire and then were entering it as uninvited guests. Here, therefore, depopulation progresses with the devastation and the decline of civilization, finding its most striking expression in the dwindling and complete disappearance of innumerable prosperous communities.

This is a summary of facts which can hardly be contested. It is therefore evident that the steady shrinkage of population and the crumbling of the Empire cannot be attributed to wars. It stands to reason that permanent desertion of entire countrysides cannot be caused by temporary devastations of war, for war cannot rob the fields of their fertility. Exhaustion of the soil, on the other hand, will lead to its desertion in time of peace and of course still more so in times of war. The "economic and social conditions," to which Meyer refers—conditions which since the later period of the Republic were increasingly depopulating Italy, Sicily and Greece, conditions that the Principate and its peace could not cure—were plain exhaustion of soil.

The inner decay to which Meyer refers, which in an era of complete peace was depopulating the Empire and its provinces, that "inner decay" which made the mighty Empire fall and crumble, that "inner decay" in all its manifold manifestations was in the last analysis entirely based upon the endless stretches of barren, sterile and abandoned fields of Italy and its provinces.

Roughly speaking, Roman political and economic life found itself involved in a series of vicious circles. It is only too well known that as long as Italian land was productive and of value, the struggle for that land was the keynote if not the very content of Rome's political struggles. The wealthy were either in lawful possession or were unlawfully using the public land, and hence they were opposed to its subdivision among colonists.

Rome was not without patriots and prophets, to whose vision it was revealed what the Fates had in store for it. They were the great land reformers—and according to tradition they all had to die. Thus, Spurius Cassius, a consul and a triumphator, is said to have been executed in 486 B. C. Marcus Manlius, Rome's greatest hero, who saved its capitol in the Gallic siege, was executed in 384 B. C. The Gracchi, though foully murdered, after a fashion succeeded; but it was already too late. As the Roman farmers were vanishing from the countryside and the farm centers, the small municipalities were losing their significance and the population of the city of Rome was increasing.

In proportion as the Roman fields were becoming exhausted, Rome had to rely upon grain from other lands. The conquest of grain-producing countries opened new rich fields of exploitation to the Roman money-men and to its statesmen with an eye on plunder. But to keep the people alive on the bread and to satisfy the appetite of the wealthy with the loot of foreign lands, great armies and a manhood superior to that of the barbarians was required.

As foreign provinces and not Italian lands became the source of Roman wealth, as the population of

Rome became too large, too motley, too complex an element to handle, then indeed even the optimates became strong advocates of colonial assignments. Thus we find the versatile Cicero, who once so eloquently opposed colonization, supporting the colonial projects of the tribune Flavius, because by such measure could "et sentinam urbis exhauriri et Italiae solitudinem frequentari" [1]—"the city might be emptied of the dregs of the populace and the deserted parts of Italy repeopled." But the colonies of the later Republic and the Principate could not be successful. It was not a surplus of a farming population that was now being settled in new colonies. Rome permitted its farming population to be wiped out, and then tried to make farmers out of idle city paupers and old army veterans. But army veterans or city rabble will not make successful farmers, even on good soil; and as a rule the land assigned to the new colonies in Italy had already ceased to be paying farming land. The colonies did anything but flourish; the colonists as a rule were quite willing to part with their allotments as soon as they could legally do so. Thus the same Cicero pointed out how some vast colonies of Sulla turned in no time into latifundia owned by a few.[2] Later on, the sources no longer report to us that the colonists were selling their allotments, they do report that they *deserted* them. So, for instance, Tacitus tells us that "the veteran soldiers entitled to their discharge from service, were settled in Tarentum and Antium so as to in-

[1] Cicero, Epistolae ad Atticum, i, 19.

[2] "At videmus, ut longinqua mittamus, agrum Praenestinum a paucis possideri." Cicero, De lege agr., ii, 28.

crease the population of these deserted localities, but many of them wandered back to the different provinces, in which they had served."[1]

On the face of it, it was a hard proposition even for good farmers fresh from the soil; for the fields assigned to them were already abandoned for good and sufficient reasons. To settle there old veterans without families and to expect them to succeed on these abandoned fields, was to expect miracles. As a rule, expected miracles fail to materialize. The countrysides remained abandoned; "Italiae vastitas" stared the contemporaries in the face; and Italy's great historians marveled how sections of Italy that in their times were almost entirely deserted could in former days send forth legion after legion of invincible warriors. The historians marveled, and surmised that these country districts were once upon a time thickly settled. Strange, but true! "Simile veri est . . . innumerabilem multitudinem liberorum capitum in eis fuisse locis, quae nunc, vix seminario exiguo militum relicto, servitia Romana ab solitudine vindicant."[2]

But perhaps no other historical document is so complete as Dio Chrysostom's story of the Eubœan Hunter. The time is the first century A. D. and the place happens to be Eubœa. There, too, the small farms first became consolidated into latifundia, and then these latifundia were abandoned.

Nearly two-thirds of our land is desolate from neglect and lack of inhabitants. I too possess a vast acreage (many plethra), like many others, I suppose, and not only in the mountains but in the

[1] Tacitus, Ann., xiv, 27.
[2] Livius, vi, 12.

valley. Should any one care to cultivate them he could not only have them rent-free, but I would gladly pay him money in addition.[1]

The local orator introduced to us by Dio goes on urging the citizens to take up the cultivation of abandoned land, for deserted land is a useless possession. Let any one cultivate as much as his capital may allow him to do, for it may save the remaining population from its two cardinal ills—idleness and poverty. The orator is suggesting that the land should be given to any one rent-free for the first ten years, and later on for a moderate tax upon the productivity of the soil, but the tenant should not pay any taxes upon his cattle. Should an alien care to take up land, he should be welcomed to do so and he should remain exempt from taxes for five years but pay for his land later double the rate of a citizen. If, however, an alien bring under cultivation 200 plethra then citizenship should be bestowed upon him as a reward so that many should aim at such an achievement.[2]

The picture of Dio is undoubtedly true to life; and we must remember that these conditions were gradually encroaching not only upon Italy and Greece but upon the other provinces with the exception of Egypt. Material which characterizes the later Empire you will find in the Church Fathers, as, for instance, Salvianus talking about Spain, of which but the name remains, of Africa that was, of Gaul that is devastated.[3] Of

[1] Dio Chrysostomos, Oratio, vii.

[2] *Ibid.*

[3] "Denique sciunt hoc Hispaniæ, quibus solum nomen relictum est, sciunt Africae, quae fuerunt, sciunt Galliae devastatae. . . ." Salvianus, De Gubernatione Dei, iv, 4.

course, Salvianus, Theodoretus in his letter to the Augusta Pulcheria,[1] Basilius, Libanius and many others put the blame on the oppressive taxes. In a sense they are right. But two points must be considered. First of all, when the productivity of the soil is very slight, even a light tax is oppressive. Secondly, it must be borne in mind that if economic conditions are preventing a portion of the population from supplying their quota for the needs of the state, the tax quota of those that have not yet been economically wiped out may have to be large. Since we know that a substantial part of the population has become so proletarized as to be a charge upon the public, some other part of the population had to bear the increased burden. The increasing tax pressure, whether relative or absolute, is in the case of Rome, therefore, obviously largely due to agricultural decline. In their turn, disproportionate taxes ruthlessly collected are quite sufficient to compel the farmer to tax the productivity of his land to its uttermost and thus to hasten the process of spoliation.

Let us now see what effect the exhaustion of the soil and the desertion of the fields had upon the body politic. For a farmer, children are a blessing. For, if a laborer is worth his hire, children are certainly worth more than their keep. In fact, lack of children is even now a hardship to a farmer; but in olden days with the much more primitive instruments of production it was a calamity. The less the productivity of labor, the greater is the effort, the greater the mass of human labor, the larger the coöperation required by a farm-

[1] Theodoretus, Epist., xliii.

ing unit. For such a coöperation, if children were lacking, men had to be bought or hired. Under certain conditions several families had to coöperate and live in a relatively large group to meet the exigencies of farming. You will find it in Greece, in the five-generation groups of the Welsh, in Slavonic Zadrugas, etc.

But that is another story. Certain it is that under wholesome circumstances, in the past as well as in the present, race-suicide is not a farmer's pastime, unless the density of population in its relation to the available land area has reached a saturation point. But what do we see in Rome already in the first century after Christ? The Roman writers marveled at Egypt and the prodigious fertility of the Egyptian race. Thus so scholarly and enlightened a writer as Columella firmly believed that to Egyptians and Libyans (northwest Africa was then still a granary) most exceptional capacity for the propagation of their kind was given. Their women, he tells us, are bearing twins every year.[1] The elder Pliny is not satisfied with twins; he insists upon Egyptian triplets, yea, even triplets did not suffice; the Egyptians seem to have gotten the habit of being born in litters. He explains the phenomenon by the woman drinking the fruitful (fetifer) water of the Nile.[2] Many other contemporary writers marveled at Egypt's great population, yet all that Egypt in matter of population can be credited with is that it held its own—seven millions—or possibly increased half a million in the course of the three centuries that

[1] Columella, iii, 8.
[2] Plin. H., N., vii, 33.

elapsed between the Ptolemys and Vespasian. Egypt was the only province whose soil could not become exhausted, because of the overflows of the Nile, and Egypt was the only province which maintained its population. This was regarded as a marvel. From the Roman point of view, a marvel it probably was. For the general depopulation had become in Rome a matter of grave concern.

In Rome birth-control and a disinclination to marriage became widespread. So Ovid tells us: "Raraque in hoc aevo quae velit esse parens." Of course the women were blamed. It is a subject that always invited loquaciousness; but we have any amount of evidence that we are dealing here, not with Rome's particular depravity, but with a phenomenon that hangs closely together with the decay of the farming population. Polybius is one of the many who discussed the subject. First of all we find that the same phenomenon is characteristic of Greek city life. The casual connections, the reasons that Polybius gives, may be of doubtful value to us, but his statements of fact are of importance. Here are two. In one Polybius tells us that in his time Greece was suffering from childlessness and general depopulation, the cities were becoming deserted and the fields were not yielding, though they have known neither war nor plague. The people became greedy idlers. They did not care any longer to marry; if they did marry they did not care to bring up more than one or two children that they might inherit the undivided fortune of their parents.[1] This is one state-

[1] Polybius, xxxvii, 9.

ment; in another Polybius tells us that the farmers constitute the most prolific portion of the population. In a certain portion of the Peloponnesus in Elis, farming was still flourishing, the countryside was thickly populated.[1] These two statements taken together certainly throw light on the situation. But there is another mistake which we must beware of: we must not paint the devil blacker than he is, and we must not regard great wealth as the sole cause of corruption, nor Rome as its seat and abode. Suppose we make a little excursion to the small provincial town of Ilion, and follow Eduard Meyer's calculations on the basis of Schliemann's Trojan excavations. We find a list of 102 citizens, enumerated with their families, 38 of whom are married, 64 not married. Of the 38 married families, 17 are childless, 21 have children. The 21 families having children have, all told, 19 sons and 12 daughters. Seventeen families have no children, 14 have one child, 4 have two children, 3 have three children. There is no family which had more than three children. Several widowed mothers live with their respective sons, eight widows live by themselves. Thirty aliens who have obtained citizenship are also mentioned in the record, of whom few have wife or child; one had his mother living with him.[2] This statistical miniature picture leaves one speechless. We have no statistical data for Rome. The situation in Rome was certainly not worse than in that romantic little town of Troy. Certain, however, it is that the degree of de-

[1] Polybius, iv, 73.
[2] Eduard Meyer, Kleine Schriften. Halle, 1910, p. 167.

population was such as to compel extraordinary measures. And so it came to pass that morals and children became political issues. Recall the puritanic campaigns of Cæsar's heir, remember the laws against adultery, and laws restricting the property rights of unmarried men and childless couples. Special privileges were granted to parents of moderate-size families—but the "jus trium liberorum" was as a matter of fact conferred even on confirmed old bachelors! It became an honorary degree! The gravity of the situation was clear to every one,[1] and hence Augustus distributed large stipends to parents of sons and daughters —one thousand sesterces for each child.[2] Later the rearing of children was encouraged in a systematic way by large special alimentary foundations, from the interest of which regular stipends were given to a large number of children both in Rome and throughout Italy.[3] The great foundations were first established by the Emperors Nerva and Trajan. Their example was followed not only by the Emperors Hadrian, Antoninus Pius, Marcus Aurelius, Alexander Severus and others, but also by private philanthropists. It became a type of philanthropy that particularly commended itself not only in Italy but also in the provinces. These foundations provided not only for boys, but for

[1] Speaking of the Censor Metellus' famous remarks about matrimony, Gellius points out: "Quod fuit rerum omnium validissimum atque verissimum, persuasit, civitatem autem salvam esse sine matrimoniorum frequentia non posse." Gellius, i, 6.

[2] "Iis, qui e plebe regionis sibi revisenti filios filiasque approbarent, singula nummorum milia pro singulis dividebat." Suetonius Augustus, xlvi.

[3] For sources see Marquardt, Römische Staatsverwaltung, Bd. ii, (2nd ed., 1884), pp. 141-147.

girls as well; and special foundations were also established for girls alone. So Antoninus Pius, in honor of his wife Faustina, established a large foundation to provide exclusively for young girls—the puellae Faustinianae.

In the beginning of the fourth century we find still more far-reaching alimentary attempts. To put an end to infanticide Emperor Constantine in 315 orders his fiscal administration to provide for the children of all poor parents residing in Italy, who have not the means to provide for and educate their children.[1]

The same Emperor Constantine issues a similar law for Africa in 322, ordering the treasury to provide for the children of poor parents, so as to put an end to their selling or pledging their own children, or, even worse, having them starve to death. For, adds Constantine: "Abhorret enim nostris moribus, ut quemquam fame confici vel ad indignum facinus prorumpere concedamus." [2] The sentiment is a noble one, but back of all these measures is the more and more pressing necessity of maintaining the Roman population if the Roman state is to be maintained.

The alimentary provisions could obviously be of value only as measures of relief. They were trying to affect the results of a certain given situation, without

[1] Cod. Theod., xi, 27, 1. "Aereis tabulis vel cerussatis aut linteis mappis scribta per omnes civitates Italiae proponatur lex, quae parentum manus a parricidio arceat votumque vertat in melius. Officiumque tuum haec cura perstringat, ut, si quis parens adferat subolem, quam pro paupertate educare non possit, nec in alimentis nec in veste inpertienda tardetur, cum educatio nascentis infantiae moras ferre non possit. Ad quam rem et fiscum nostrum et rem privatam indiscreta iussimus praebere obsequia."

[2] Cod. Theod., xi, 27, 2.

in the least affecting that situation itself. This probably was clear to the Roman administration. Thus we witness under the Principate almost frantic attempts to create a farmer class and to repopulate the country districts. Cæsar and Augustus made the greatest efforts in that direction. Augustus established not less than twenty-eight colonies. His successors followed in his footsteps. Nerva went so far as to spend the vast sum of 60 million sesterces in purchasing private latifundia in Italy and dividing them among colonists.

Soon we see the government attempting much more drastic measures; we are confronted by legislation as unprecedented as it was amazing from the legal point of view. But these measures had become matters of course and of necessity probably long before Pertinax. For Pertinax's short reign they happen to be recorded. So Herodian tells us that throughout Italy and in the provinces every one was permitted to take possession of waste land and abandoned fields, even if they were the Emperor's own property, and whosoever was tilling those fields was to become the rightful owner and proprietor of the soil.[1]

Dio Chrysostom's flights of oratory have actually become stern realities. There can be little doubt that the agrarian situation throughout the Empire was very much like the one in Eubœa described by Dio. Nor can there be doubt as to the cause for the desertion of the fields.

If this actual situation is kept in mind, the agrarian legislation as embodied in the Theodosian and Jus-

[1] Herodian, ii, 4, 6.

tinian codes begins to have a stern meaning. Decrees of that type must have been issued by both Hadrian and Trajan, judging from the inscriptions of Ara legis Hadrianae and Henschir Mettich.[1] It goes without saying that the situation did not improve in the following period. Whosoever wants land, waste deserted land, is cordially welcome to it. "Quicunque possidere loca et desertis voluerint, triennii immunitate potiantur." [2] Still more explicit is the law of Valentinian, Arcadius and Theodosius [3] which first of all encourages every one to cultivate abandoned fields or fields that have been without cultivation for a long time. The original rightful owner of such fields is to have but two years in which he may claim back his property, fully compensating the new occupant for all his outlays. After a period of two years all the property rights of the former owner are to be extinguished in favor of the new occupant.

Such legislation is conceivable only with a background of endless stretches of abandoned and untilled land. But the law was not successful, since the encouragement was not sufficient to induce voluntary reclamation of abandoned farm-land. The reclamation

[1] Ludwig Mitteis, Zur Geschichte der Erbpacht im Altertum. Abhandlungen der Königl. Sachsichen Gesellschaft der Wissenschaften Philologisch-Historische Classe, Bd. xx, pp. 28-33.

[2] Cod. Theod. v, ii, 8.

[3] C. J., xi, 59, 8; Cod. Theod. v, 11, 12. "Qui agros domino cessante desertos vel longe positos vel in finitimis ad privatum pariter publicumque conpendium excolere festinat, voluntate suae nostrum noverit adesse responsum: ita tamen, ut si vacanti ac destituto solo novus cultor insederit ac vetus dominus intra biennium eadem ad suum ius voluerit revocare, restitutis primitus quae expensa constiterit facultatem loci proprii consequatur, nam si biennii fuerit tempus emensum, omni possessionis et dominii carebit iure qui siluit."

of barren land is thereupon ordered as an obligation upon every possessor of estates under cultivation. He is to cultivate the barren and waste land within his estate. Nor can the possessors of estates sell their land under cultivation without at the same time disposing of the barren and unprofitable parts of their estate which the purchaser is to cultivate. The legislation was known as the 'επιβολή or "*iunctio,*" the "imposition of desert to fertile land."

Let us face now the problem of the so-called origin of the colonate in Europe. The problem has been much discussed, for it is of vital importance, not merely as a detail in Roman history but as a tableau that introduces us into the Middle Ages. It is the problem of the origin of medieval serfdom. Instead of arguing with the many different conceptions of the origin of the colonate, let us conceive its origin as intimately correlated with the land legislation and with the legislation regarding slaves. If so, all these correlated problems are state problems relating to the unprofitableness of farming as a universal social phenomenon. And if so conceived the very text of the law suggests the answer and leaves no problem to be solved.

In the introduction to the eightieth novella of Justinian, we are told that provinces are gradually becoming depopulated and that our great city is burdened by all sorts of people, especially farmers, who have abandoned their homes and deserted their fields. Why they were deserting their fields was obvious. What kind of people were living and working on the fields? They had all kinds of names, but there were but two different varieties: slaves and freemen. The slaves were trained agricultural slaves; the freemen were tenant farmers. When agriculture became very unproductive and the owner of the estate got no profit from his slaves and his estate, his natural tendency was to dispose of his slaves otherwise. The interests of the state were opposed to such tendency. The remnant of the agricultural population had to be saved. Hence the law prohibits the removal of the agricultural slave from the soil. It improves his condition. He is decreed to be attached to the soil; he may not be sold without the soil; he may marry; he may not be separated from his family. The purpose in this cannot be doubted; but such enactment does not yet keep the entire agricultural population on the soil. There is another large category—the free tenant. His freedom was never before questioned, but the legislator is not in a mood to bother about niceties of the law. The free tenant is deserting the fields rapidly. To prevent this he too is being bound to the soil, he becomes a serf. The date of the enactment is lacking, as is the original decree, but there are numerous decrees against fugitive slaves and coloni,[1] and special drastic legislation against tricky evasions of the law. Notice especially the following decree:

Just as in the case of persons bound to the soil by birth (*originarii*), so in that of slaves settled on agricultural land and listed as liable to the poll-tax, it is absolutely forbidden that they be sold off the land. Nor by tricky misconstruction shall the law be so evaded, as has repeatedly been done in the case of *originarii*, that

[1] C. J., xi, 48, 6; *ibid.*, xi, 48, 11.

an *entire estate shall be deprived of tillage* by transferring a small portion thereof to the purchaser of the slaves. On the contrary, whenever an entire estate or a definite portion thereof comes into any one's hands, there shall go with it just so many slaves and *originarii* as were settled on such entire estate or portion thereof in the time of its former owners and possessors; and the purchaser may regard the price paid as money lost, since in spite of the sale, action will lie on the part of the vendor to recover the slaves and also their children born after the sale. And if for any reason the vendor shall have neglected to enforce his legal right and shall have died without bringing suit, we give the action for recovery both in favor of his heirs and against the heirs of the purchaser, depriving the latter of the plea of prescription by lapse of time; for no one is to doubt that he who buys anything against express statutory prohibitions is a dishonest possessor.[1]

This decree is important not alone because it recognizes no time limitation and gives action of recovery even to the heirs of the seller against the heirs of the purchaser—an unusually drastic measure. It is important also because it tells the story: it explains the reason why.

Here is obviously a common evasion. Why a common evasion? Because the evasion is lucrative. What is the policy of the state? It is to maintain agriculture even against the interest of the owner of the slaves.

Here you have the *raison d'être* and origin of the Roman colonate. It is fundamentally the very same reason that led to the emphyteusis, the ἐπιβολή-the same reason that inspired the entire agrarian legislation of the doomed empire.

There are endless details relating to this as to any other subject; they are outside of our scope. For our

[1] C. J., xi, 48, 7.

set task was not to write about Rome's life, but about Rome's end. There is, of course, abundant material that we cannot touch upon here.

For instance, the material relating to early Christianity in its popular acceptance, through its spiritual reflections of existing conditions, furthers understanding. For the road that started at Golgotha led to Rome and through the Roman Empire. What were the aspirations of those who traveled it? Did they in no way reflect the conditions of the time? What do theologians tell us about the apostolic age? What is its peculiar characteristic? It is recognized and unquestioned that early Christianity was an end-of-the-world religion.

So writes Cyprian to Demetrianus:

> You have said that all these things are caused by us, and that to us ought to be attributed the misfortunes wherewith the world is now shaken and distressed, because we do not worship your gods. And in this behalf, since you are ignorant of divine knowledge, and a stranger to the truth, you must in the first place know this, that the world has now grown old, and stands no longer in its pristine strength; nor has it that vigor and force which it formerly possessed. This, even were we silent, and if we alleged no proofs from the sacred Scriptures and from the divine declarations, the world itself is now announcing, and bearing witness to its decline by the testimony of its failing estate.[1]

That much we have perhaps heard before, but here is a new note: "Although the vine should fail and the olive deceive and the grass languish with drought on the parched field, what is this to Christians?" [2]

Perhaps even more expressive is somewhere in

[1] Cyprianus, to Demetrianus, ii.
[2] *Ibid.*, xi.

Lactantius this sentence: "There will be a dreadful and detestable time in which no one would choose to live. In fine, such will be the condition of things that lamentations will follow the living and congratulations the dead."

Quite adequate illustrations to this text you may find in the pictures of misery and desolation in Salvian's "De gubernatione Dei." Is not the material basis to be found in such conditions for celibacy, asceticism, monasticism? And were not the same conditions after all responsible for Roman childlessness, though quite unaccompanied by mortification of the flesh? It may be difficult for us to understand an atmosphere of social and political doom. Let us go back then to another people, to another city, that were about to be destroyed by Rome. Jesus was carrying the cross, followed by lamenting men and women:

> But Jesus turning unto them said: Daughters of Jerusalem, weep not for me, but weep for yourselves and your children. For behold, the days are coming, in which they shall say, Blessed are the barren, and the wombs that never bare, and the breasts that never gave suck. Then shall they begin to say to the mountains, Fall on us; and to the hills, Cover us.[1]

Here, however, we are touching upon a subject which should be treated by itself.

It is claimed that there is but one understanding; the misunderstandings are legion. To guard against misunderstandings is impossible. Yet I know that many a charitable reader will sympathetically suggest that while the exhaustion of Roman soil was an im-

[1] Luke 23, 28-30.

portant factor I can hardly mean to insist that it was *the* sole factor responsible for Roman decline and fall. For it is not credible that so rich and so complex a texture of life should depend upon any single factor.

Such would not be my assertion, nor is it my attempt. I have not undertaken to explain the complex fabric of Roman life; we are dealing here with the relatively simple problem of its disintegration. All that this study shows is that the progressive exhaustion of the soil was quite sufficient to doom Rome, as lack of oxygen in the air would doom the strongest living being. His moral or immoral character, his strength or his weakness, his genius or his mental defects, would not affect the circumstances of his death: he would have lived had he had oxygen; he died because he had none. But it must be remembered that while the presence of oxygen *does not explain his life*, the absence of it is sufficient to explain his death.

There is one other misunderstanding which I should like to guard against. So far as argumentation is concerned, this essay might be considered a continuation to the study published some time ago, dealing with the medieval village community.[1] The reader will find there this statement:

> Go to the ruins of ancient and rich civilizations in Asia Minor, northern Africa or elsewhere. Look at the unpeopled valleys, at the dead and buried cities, and you can decipher there the promise and the prophecy that the law of soil exhaustion held in store for all of us. It is but the story of an abandoned farm on a gigantic scale. Depleted of humus by constant cropping, land

[1] Simkhovitch; "Hay and History." Political Science Quarterly, vol. xxviii, pp. 385-403, September, 1913.

could no longer reward labor and support life; so the people abandoned it. Deserted, it became a desert; the light soil was washed by the rain and blown around by shifting winds.[1]

I should hate to be responsible for a new fetish, an interpretation of historical life through exhaustion of soil. It is silly.

First of all deeply and gratefully is it felt that life with all its pain and its glory can be lived; word or brush may aspire toward its all too inadequate expression, but never will the scholar methodically and mechanically figure it out and interpret it.

But it is a mistake to think that social science is dealing with life. It is not. It deals with the *background* of life. It deals with common things, with what lives had in common, common conditions of existence, common purposes that these conditions suggest. *They* can and must be scientifically explained and determined, if social science is to be taken seriously. Scientific determination is accurate determination. What forces that circumscribe and govern our life must we unquestionably accept? Obviously, the physical forces. Under certain conditions we are born, we live and die. The limits of our mortal existence we cannot transgress. Nor can we change the heavenly course of suns and planets; we do not govern the seasons of the year; they regulate our life.

Within the laws of nature our lives begin and end. They limit and compass our existence.[2] But the laws of nature without our active participation do neither

[1] Simkhovitch, *op. cit.*, p. 400.
[2] With apologies to Goethe's "Nach ewigen ehernen grossen Gesetzen müssen wir alle unseres Daseins Kreise vollenden."

feed nor clothe us. This active participation we call our work, our labor. Social labor varies in its productivity. At all times this productivity had and has its limits. *These limits of the productivity of our labor become, for society, physical conditions of existence.* Within these limits our entire social life must move. These limits life must accept as mandatory and implacable; to them it must adjust itself.

The history of the productivity of our labor is the foundation of a scientific economic history, and the backbone of any and all history. Every law, every statute, every institution has obviously some purpose. But how are we to understand the purposes of the past unless we know the conditions which those purposes were to meet? The accurate knowledge of the productivity of our labor can explain to us why things were as they were, why they became what they are and what one may expect from the future.

In this study, however, which is not concerned with the details of Rome's life, one single, major and strikingly variable productivity factor suffices to solve the problem. That factor—the exhaustion of Roman soil and the devastation of Roman provinces—sheds enough light for us to behold the dread outlines of its doom.

HAY AND HISTORY

WHAT is here to be discussed concretely is the village community. This has already been, by actual count, considered in more than a thousand books and essays, and written about under titles more seemly and more modest. Yet I venture to tax the patience of the reader with the old story once more; not even with the story, because all I am offering here is an interpretation.

I ask only this question: Can the reader tell me *why* the village community was so prevalent in Europe, regardless of race and clime? The reader, if he belong to one school, will say that the village community is a survival of an early Teutonic or even of a universal ancient custom. If he belong to another school, he will cautiously answer that wherever we meet this institution in the past it is invariably in connection with the manorial system. Thus the thousand contributions to the subject do not answer the question *why*. Those of one school practically say that it existed but they do not know why; those of the other school intimate that the lords of the manor introduced it but that they do not know why. Worse yet, not only has the question *why* not been answered; it has not even been asked.

Hence there is after all a justification for one more essay on the subject, in which the simple question is asked and a plain answer given.

This is an interpretation, not a story, and it is possible that some kindly reader may come to the conclusion that after all the writer is more interested in a problem of which the village community is but a manifestation,—the problem of the life and death of the land and the peoples thereon. Possibly so. But let us be concrete, let us talk village community.

First of all, what have the discussions been about? The controversy covers several problems: did or did not the early Teutons have a "Mark Association"; is or is not the village community of to-day a survival of a system when all land was universally held in common ownership; do we find the *Mark Genossenschaft,* the community of free Mark associates, in England, Scandinavia, and among the Slavs; is or is not the village community in Germany, Russia, France, England, elsewhere, after all but a product of serfdom and the manorial system; is or is not the Roman *villa* responsible for the manorial organization of western Europe?

These are the main problems that have been discussed for many decades under the general heading of "village community." The controversies have been too often acrimonious. While they have unquestionably added to the sum total of our knowledge about miscellaneous historical subjects of importance, the real problem has been obscured rather than illuminated by the wealth of historical data.

Let us see what is the real problem that a village community presents. There are thousands of old villages in Europe, where the holdings of the individual peasants have never been inclosed. The holdings of one are intermixed with the holdings of his neighbors in the open fields. Let us examine such a village.

First of all, we find all the homesteads grouped together. In some localities they form one long street; in others, two streets; in still others, they are laid out in a semi-circle. Near the homesteads are the barns, stables, hovels, vegetable gardens and a few fruit trees, but never the field that belongs to the individual farmer.

The farming land presents a curious sight. It looks like a patchwork quilt. You see a number of land areas, of flats, of plots, as a rule square or oblong in shape, each of them divided into very numerous, narrow strips of land. Sometimes the three fields of the village form three quite uniform flats, subdivided into numerous strips, long and narrow, running in the same direction; but oftener you will find that each of the three fields is made up of many such flats, each of them subdivided into strips or ribbons of land. These strips belong to different owners, but they are not fenced. They are separated from each other by balks of turf, or unplowed land. Strips in some patches are, at times, so narrow that one wonders how they could be cultivated.

The individual peasant may own ten, twenty, or more strips in each of the three farming fields, but the strips of the individual farmer, even in the same field, do not adjoin each other. They are at times quite a distance apart. They are the farmer's individual property which he has inherited and which he may sell. But in the use of his property he is necessarily restricted. First of all he is restricted in the rotation of crops, and type of tillage. The third field is the fallow field on which the cattle of all the villagers pasture. The cattle are also pastured on the stubble as soon as the harvest is removed. Hence it becomes necessary for all the members of the community not only to sow the same crop on the same field, but to sow at the same time and to harvest at the same time.

Besides the cultivated fields the village as a rule has some waste that serves as a permanent pasture, woods, and a meadow. None of this is subdivided or fenced. It is used in common, with restrictions varying in different localities. As a rule with a definite acreage in the cultivated fields goes a proportionate and quite definite acreage in the meadows. Let us say that a certain farmer owns two acres of meadow land. It is left, however, to a yearly drawing of lots to determine where his two acres are to be located. He will get neither more nor less than the two acres to which he has a title, but the location of these two acres may vary from year to year. As soon as the meadow hay is mowed, the entire meadow is thrown open for common pasturage of all members of the village community. Sometimes all varieties of cattle are allowed there; oftener sheep will be restricted to the waste and fallow fields or to a special part of the meadow.

This is the organization of the village community as we find it still in numberless localities of northern, western, and southeastern Europe, an institution that has given rise to so prodigious a literature.

It is not an idle problem, either. Agriculture was

until recently the sole basis of state and society. It is, and will remain, of paramount significance. Anything so fundamentally characteristic as is the village community of European farming is of fundamental economic and historical importance. The problem of the village community is not a new one. The enclosure of the commons shook the very foundations of sixteenth-century England. Yet curiously enough it became a scientific and bookish problem only in the nineteenth century. This bookish spirit is nowhere so well expressed as in Wagner's rejoinder to Faust:

> I often had myself fantastic notions,
> But never have I felt the like emotions.
> 'Tis tiresome on green woods and fields to look,
> The bird's wing crave I not in slightest measure.
> How otherwise bears us the mental pleasure
> From page to page, from book to book!
> Then grow the winter nights so lovely fair,
> A warm and blissful life all limbs pervading,
> And oh! unroll'st thou e'en an ancient parchment rare,
> All Heaven descends to thee that knows no fading.

Tiresome as it may be to look at green woods and fields, let us do so for a change. If we should take with us a plain American farmer and show him a European village community he could not possibly believe his eyes. First of all he would observe the homestead with the farm buildings all clustered together, far away from the farming land. This would naturally look to him just as if an American farming community should live in a city, keeping there the horses, fodder, *etc.*, but going out every morning to farm somewhere in the country. The American farmer would hardly know what to think about such a situation. True, he knows of wealthy farmers who live in the town in the winter, or even go visiting or traveling in the winter, but they can do so because during the season they are on their jobs without wasting a minute of their own or of their horses' time. What are we to tell this plain American farmer? We consult all the books and find that they are unanimous on this point. In fact it is the only point on which the various writers are unanimous. The European peasant lives in villages *for protection.* Protection of what? The farm land, the crops, the cattle on the pastures? Of course not. They live in villages for their own protection, for the protection of themselves. Shall we give this answer to the sturdy son of an American pioneer? His comment could easily be guessed, as well as his praise and thanksgiving for having come from a different stock. But that is where our farmer is quite mistaken. His stock, whether English or Irish, German or Scandinavian, Magyar or Slav, have all lived in precisely such villages in the old country.

But, curiously, if we consult documentary evidence we shall find that early agricultural pioneers and settlers in newly colonized European territory, where they needed "protection" the most, lived as a matter of fact not in villages but on single farms. Their "villages" consisted of single homesteads, but as the population grew even these communities often developed the same uniform type as the European village, with all the homesteads huddled together. Instead, however, of consulting thousands of documents let us try another method. Let us give our poor ancestors the benefit of the doubt and assume that they had some common sense. If they so universally persisted in holding to an arrangement so obviously and seriously inconvenient there must have been some good reason for it. When we proceed on this assumption, we almost invariably find that a "good reason" is a plain, technical, and economic necessity. So it is in the case of the European village. Where would you locate the farmhouses, if not together in a special area? Would you put the house and garden on the farm land? If so, to which of the farmer's numerous and widely scattered strips and ribbons of land would you attach his homestead? And if his house were securely attached to one of his strips, would he not be just as far away from all his other strips as when he lives in the village? Furthermore, if his homestead and garden should be located on one of his strips in one of the three farming fields, he would find himself every third year on a fallow field, where the cattle of all his neighbors are grazing, and perhaps miles away from the strips he is cultivating in the two other fields; besides his buildings and his garden would interfere every year with pasturing on the stubble.

Thus the village community and the possession of isolated, intermixed strips of land necessitate living together outside the area given over to tillage, and thus there is formed a village street.

Now let us turn to the farming land.

The reader, of course, knows that in no occupation is the time element of so decisive importance as in farming. If the land is not ploughed and the crop is not planted at the right time, the partial loss, and at times the total ruin, of a crop may be confidently expected. This is particularly the case on heavy clay land, which drains very slowly. If you do not happen to plough the land and plant your crop at an early favorable moment, a single rain may delay your work for some time, and just as the ground is about fit for work another rain will emphasize the point that farming brooks no delay. Yet if the farmer should be driven in despair to plough and to sow in the wet he would have good cause to remember one of Tusser's "Five Hundred Points of Good Husbandry,"—that "who soweth in rain he shall reap it with tears."

Thus farming is in a sense a race with the season, with the weather as a constantly menacing and incalculable element. That is why from time immemorial the farming season has taxed the effort of men and beasts to the very limit of their physical endurance.

Now let us have another look at the village community.

Let us take an English example: South Luffenham in Rutland was not inclosed until 1879. It had 1074 acres, divided into 1238 pieces or strips, among twenty-two owners. Think of a farmer who has to attend forty-odd acres in nearly sixty scattered pieces, dotting a surface of over 1,000 acres! Think of the waste of time and effort in cultivating such land strips! But this happens to be a small village community. Imagine a large community, where the acreage of the individual farmer is scattered in intermixed strips over an area of many thousand acres! Why should humanity have so handicapped itself intentionally?

It is a survival, you will say. Quite true. One finds the situation in the Middle Ages whether the farmer was freeman or serf. But does the word "survival" in this particular case explain why there was a survival? Anything that is serviceable or even indifferent may survive. But is it sensible to assume that a most serious and burdensome handicap can survive without sufficient reason for surviving? But perhaps you do not belong to the school which believes that the village community is a survival from the dawn of civilization; perhaps you accept the only possible alternative, that the village community is a product of the manorial system?

It is clear that the intermixed strips of land result in waste of energy of both men and beasts. If the man is a serf, the loss of his time and the impairment of his efficiency is in the last analysis the loss of the lord. The serf must be provided with a living from the product of his labor and only the surplus over this goes to the landlord. The lack of productivity is therefore the landlord's loss. Why then should the landlords have introduced such a system in England as well as in Roumania, in Germany as well as in Russia, in France as well as in Hungary? How can such uniformity be explained? Manorial systems varied from age to age and from locality to locality. Here they commuted services to money rent, there the landlord would not even reserve a domain but left everything to the peasants for an "obrok," a heavy rental paid in produce. They have varied the internal workings of the village community. In Russia, for instance, they introduced periodical redivisions of the land-allotments of the serfs. The serfs there did not have a virgate-system. Their holdings varied and were dependent upon the fluctuations in the size of the individual serf families, upon the increase of the entire serf population of the manor, and upon the acreage of the manor that could be set aside for the serfs. One could go not only from country to country but from county to county, yes even from manor to manor, and write a story of greater or lesser variations in agrarian relationships, but the underlying basis was the same village community! There were times and localities in which the serf had no claims or rights sanctioned by custom, where the landlord's arbitrary power could not be disputed; and yet the village community was there! How then is the village community to be explained?

We have seen that the village community was a time-robbing and in many other respects a handicapping institution. Already in the sixteenth century Fitzherbert and Tusser were heaping curses upon the village community. Discussing this system Tusser says:

> What drudgerie more anywhere?
> Lesse good thereof where can you tell?

Why then did such an institution persist in surviving? There must have been some circumstance either of a compelling or of a compensating nature. Where are we to look for it if not in the prevailing methods of farming, the treatment of the soil, and the crops that were cultivated?

There are some very suggestive lines in Lucretius. At the end of the second book where he discourses about the world's growing old, he says about the earth:

> She brought forth herbs, which now the feeble soil
> Can scarce afford to all our pain and toil.
> We labor, sweat, and yet by all this strife
> Can scarce get corn and wine enough for life.
> Our men and oxen groan and never cease,
> So fast our labors grow, our fruits decrease.
> Nay oft the farmers with a sigh complain
> That they have labor'd all the year in vain. . . .

But did not people know about improving the soil with manure, and making it productive? Yes, indeed, they knew all about it and practiced manuring. But to have manure for the improvement of the soil one has to keep cattle. Did they not keep cattle? Yes, indeed, they kept cattle; but the question is, could the individual farmer keep on his land *enough* cattle to improve or to maintain the productivity of his entire farm? With the cattle kept upon his entire farm he could undoubtedly have fertilizer enough to maintain the productivity of at least a certain portion, but if he should apply manure to a part only, the rest of the land would rapidly deteriorate. If he chose rather to distribute an insufficient amount over his entire farm, the whole would gradually be depleted of nitrogen, humus, *etc.*, and steadily lose in productivity.

Let us examine the medieval situation.

The field crops were: first year, wheat or rye; second, oats or barley or beans; third, fallow. Where in this schedule does grass-seeding come and where are the hayfields? *There were none.* The economist who knows so much about the industrial revolution has overlooked another revolution that is of fully as great importance, *a revolution that fundamentally changed the basis of agriculture, that abolished the law of diminishing returns as expressed by Lucretius, later discovered by the economists (in its original version) just at the time when it ceased to be true. This great revolution was the introduction of grass-seed and of the "great trefoils," the various clovers, including later on Lucerne or alfalfa.*

There were no hayfields, therefore, before the latter half of the seventeenth century. How then could they keep cattle? They had straw and meadow grass. Meadow grass could grow only in very definite places on low and moist land that followed as a rule the course of a stream. This gave the meadow a monopolistic value, which it lost after the introduction of grass and clover in the rotation of crops. That explains why we find in Thorold Rogers' work that farming land in the last six centuries increased in value sixty times, and meadow land but twelve times. It simply means that six centuries ago meadow land was, in comparison with land under tillage, five times as valuable, relatively, as it is to-day; and it further emphasizes how indispensable a little bit of meadow was to a farm.

The meadow was used first of all for mowing, and after the hay crop was gathered it was used for pasturage. The right of pasturage was as a rule limited to the number of cattle that the commoner could keep through the winter. It is the German *Ueberwinterungsmassstab* which we find prevailing in England and elsewhere. There was no necessity for stinting the common pasturage any more definitely. The stern sway of the winter scarcity attended to that. The quantity of meadow hay gathered was a mere trifle, because the proportion of meadow land to the tillage fields was slight. So, for instance, in Hewsted, Suffolk, in 1285, on seven manors together that proportion was but 1 to 24.

We must remember that even turnips, which were known as garden roots to Tusser, were not introduced as a field crop before the end of the seventeenth century. Thus the wintering of cattle was a most precarious enterprise. Only an extremely limited number of cattle could be kept barely alive through the winter. The value of cows as milk producers was slight, even in the summer; in the winter it was trifling. Hence the saying:

> A swine doth sooner than a cowe
> Bring an ox to the plough.

Because of scarcity of milk in the winter, the farmer aimed, as a rule, to have his cows fresh late in the fall, yet four-fifths of the total annual yield of milk was in the months from April to September, when the cows were on pasture. But even then Walter of Henley, the thirteenth-century bailiff, who shows a distinct tendency to exaggerate rather than to underestimate, expects but three and one-half pounds of butter from three cows in one week. Only a good-for-nothing farmer to-day will keep a cow that is not doing better than Walter's three cows put together. But on the other hand, nothing could be expected from cows kept during the winter almost entirely on straw and tree loppings. Even straw was scarce, because of the exhaustion of the soil.[1] Barn manure was therefore obviously produced in inadequate quantities. The impossibility of getting a sufficient amount of manure for the fields is adequately indicated by the miserable pettiness of English manorial regulations, which required the sheep of the commoners to be folded on the domain of the lord.

It is not within our province to go into agrotechnical details and describe what the medieval farmer knew but seldom practised, for lack of time and because of poor means of communication, in the way of liming sour clay ground, *etc.* Plant production is determined by that one of the necessary elements which is available in the least quantity. It is a matter of record that the medieval farmer had not enough and could not have quite enough manure to *maintain* the productivity of the soil. To improve its productivity was out of the question.

If any good farmer should to-day take up a worn-out farm, which still has some fair fields, but others that have grown wild, and still others that are bare, washed and gullied, he should be able to make all the

[1] In the years 1243-8 the average yield of wheat at Combe, Oxfordshire, was 5 bushels per acre, of barley a little over 5, oats 7. The Pipe Roll of the Bishopric of Winchester gives an average yield of wheat over a large area in 1298-9 at 4.3 bushels per acre. Curtler, "Short History of English Agriculture," p. 33.

fields look alike in a very few years. As fast as fodder can be produced stock will be added, the poor lands heavily manured, farmed around and turned perhaps into an alfalfa field that will rejuvenate the soil.

Compare again the sentiment of Lucretius, "So fast our labors grow, our fruits decrease," with a statement in a recent Bulletin of the U. S. Department of Agriculture about the introduction of sweet clover—*Melolitus alba*—in Kentucky.

A decade ago many farms were coming to be abandoned, owing to their low productive capacity. Many of the fields contained gullies, which washed, making the farms even less valuable. Sweet clover was introduced as a bee plant in some of the waste places in this section and proved so efficient as a soil-improver that it has been largely utilized on a great many farms in this section. . . . As a result of the introduction of sweet clover many of these farms are no longer abandoned but are producing satisfactory incomes for those who are working them. The fields are utilized as pastures and for hay while the soil is being built up and the gullies stopped from washing. When the process of natural reclamation has gone far enough the ground is plowed and put into corn.[1]

It is only for the sake of the concrete example given in the *Bulletin*, picked up by chance, that I mention sweet clover, which is, as a matter of fact, an inferior fodder plant and far inferior to alfalfa as a soil renovator. It only shows that improvement of poor soil can be and is undertaken as a matter of course and of daily routine by every civilized farmer in modern times. The attitude towards a variety of soils of different qualities is quite different in modern times from what it used to be. In the Middle Ages, or rather in

[1] U. S. Dept. of Agriculture Bulletin 485, pp. 34-35.

the dark ages of agriculture, *the state in which the land was found was by and large the upper limit of its productive capacity. It might deteriorate, it was likely to deteriorate, but it was not likely to be improved.* Let us see what is the result of such an attitude.

If two of us chance to be joint heirs to a two-hundred-acre farm containing fields and patches of ever so many different qualities, what are we likely to do? In our day and generation we will divide it in two more or less even parts, and call it fair. I have before me some documents that show us that that was not the way an equitable division of an ancestral farm was carried through in the past.

Let us go to a region where the existence of a village community was out of the question. It is in the northernmost Russia.[1] The settlements there were single isolated farms (German, *Einzelhof,* Latin, *Mansus*). They called these farms "villages." Later on, when they actually had villages with several farms in each, they would still call their farms their "village," situated in village so and so. So, for instance, in a certain district which numbered seventeen villages in 1651, eight of these villages still consisted of one farm each, five of two farms, one of three, three of four farms.

The farms were private property, which could be bought and sold, inherited, mortgaged or disposed of by will and testament. It is important to observe what

[1] Detailed information, documents and authorities are given in the first few chapters of my book, "Die Feldgemeinschaft in Russland," Jena, 1898. A mass of facts and information is there, but I passed the real problem without ever noticing it.

occurs when they are divided among heirs, when the "village" that consisted of one farm becomes a village of two, three or more farms. Both wills and tax-books contain complete descriptions of the farm lands. And what do we find there? Was a farm when divided between two heirs divided into two parts? Not at all. Here is an example: "Village" Novinki (Newlands) divided into ten different plots, each of these plots divided evenly between the two heirs. Another farm (Village Towra) is divided according to quality into twenty-two such plots, and each of these plots into two shares.

I need hardly point out that the moment the land strips of one owner are intermixed with the land strips of his neighbors, we have before us a village community. One could not have a right of way to the various strips, nor pasture on them when they are lying fallow or in stubble, without a uniform—which means compulsory—rotation of crops.

The principle of equity that prompted such a division of inherited property is obvious. Would some eastern tribesman who left behind him four sons, four camels, four horses, four sheep and four dogs, leave to one of his sons four camels and to the other four dogs? Or are his heirs likely to receive share and share alike the different kinds of property left by their father? Sufficient reason has been given for showing why in the dark ages of agriculture a meadow was altogether a different type of property from an arable field, and why under the agrotechnical conditions then prevailing arable fields of different quality were considered a different and a *permanently* different kind of property, equitably divisible only in strips.

A small garden plot could be improved, but poor farming land had no chance of betterment. If its condition altered, it would only be for the worse. Under these circumstances the equitable division of a farm made up of pieces of land of different qualities led inevitably to the strip system.

Does the manorial system present to us a situation so vastly different? Whether we find serfs or customary tenants, whether the services are commuted or not, the obligations resting on those who were tilling the ground were as a rule uniform. One serf was required to perform the same services as another. If the services were commuted the rents and fines were as a rule uniform, a uniformity sanctioned by custom if not by law. Under these circumstances it was obvious that serfs and tenants respectively had to be provided with land of equal value.

Under agricultural conditions as they existed there was no way of getting around or of ignoring the different qualities of soil. It was like mining: some soil was pay dirt, other soil was not. Soil could not be improved nor could hayfields be created, as we create them to-day by throwing broadcast anywhere we please some seeds of timothy or clover. They were necessarily located in a certain and given place near the brook. Equitable distribution meant access to a share in the meadow, meant a division of fields into strips that went through good and bad, dry and wet soil, with rights of pasturage, of mast and firewood, *etc.;* meant, in short, a village community.

Was the village community therefore necessary and inevitable, in spite of all its inherent annoyances, which Tusser characterized as "too much to be borne"? It is an idle question. The village community was dictated neither by fate nor by fetich. There were for all we know local conditions that did not favor the origin or spread of the village community, but we do know and understand the situation which suggested the village community as the sole prevailing method of equitable distribution. It is no more a peculiarity of the Slav and Teuton or the Aryan race in general, than are clay and sand, stony hills and rich meadows peculiarities of the *"Christlich germanische Cultur."* Nor should too much stress be laid on the "equitable" side of distribution. If large estates were operated by slaves fed by the master or by free laborers paid wages in money or in produce, there would have been no village community under the manorial system. But at a given time, under the prevailing conditions, it was found more convenient to let labor maintain itself. For that purpose land had to be set aside for the laborers and then indeed the question of equitable distribution of that land had to arise. The solution was the village community.

The situation was similar when small heirship properties required equitable division. Some historians and sociologists find peculiar inner satisfaction whenever evidence shows that drawing of lots was resorted to. The drawing of lots is to them a character certificate of the "primitiveness" of the institution, it proves to them that the village community is a "survival" from the times when all land whether under tillage or not was held in common by the free associates of the "mark," or its English, Irish, Russian, Indian, *etc.*, equivalent.

Other scholars, while acknowledging the reallotment of meadow land, frown severely at any possible allotment of arable land. But they are wrong. Suppose four heirs inherit four horses, seemingly equally good, but two of the heirs seem both to have a strong predilection for the same horse. Will they be doomed to serve as exhibits in the survival gallery if they match pennies for the possession of the horse in question? Contention is never profitable and drawing lots for land strips is as good a way of avoiding difficulties as any, Certainly it is a survival, but so is common sense. For the custom of a *periodical* reallotment of land under tillage, only Russia can be pointed to with assurance. There reallotment was readjustment, because the Russian manorial system did not grant to each peasant family a "virgate" or "hufe" or a certain customary unit of land for customary services or rents, but divided the entire manor or part thereof among the peasant families in proportion to the working capacity of these respective families. Therefore, as the working capacity of the individual families increased, the land of the manor had to be redivided and the individual family holdings proportionately readjusted. Such is the simple explanation of periodical reallotment. The practice, moreover, is one of late origin. Among some of the crown peasants it was as a matter of fact introduced as late as 1830-1840, and introduced practically by force, in spite of the bitter opposition of the landholding peasantry.

I know of places in northern Russia where in the seventeenth century, land-strips, though intermixed and under the usual restrictions of a village community, but in private ownership, were seemingly redivided among the several owners. But this was because of disputes over the original division due, as a rule, to faulty surveying. They had no trained surveyors with modern instruments. When such contentions became a nuisance, the contending parties as a rule signed a document in which they expressed complete satisfaction with their strips and boundaries and agreed for themselves, their heirs and assignees, never to demand another redivision or adjustment.

The yearly reallotment of meadow land was a very widespread custom in Europe. As a rule the commoner owned a definite amount of the meadow, let us say an acre, or two acres, as the case might be. His acreage in the meadow was always proportionate to his acreage under tillage, but just which specific acreage of the meadow, which plot of the meadow, he was to mow in the given year was left to lot. We know that in the absence of herbage and clover, meadow land was of the highest value. But since, after mowing, it was used for common pasture, permanent fences of individual meadow plots were out of the question. Furthermore, permanent ownership of a specific portion of the meadow had other difficulties. The site of such a meadow we can easily imagine. It was in a valley near a stream. The land as a rule was made land, which had washed down from the hills or higher grounds. Every one knows how wayward a meadow stream is. It is prone to wander from year to year; here it abandons its course, for reasons entirely of its own choosing; there it makes a nice little lagoon, or playfully deposits some very fine gravel on what might have been good meadow grass. Every few years such a stream is doing some surveying entirely of its own. It will take a few rods of land from one farmer by running through his land and give a few rods to another farmer by adding to his land the course that it abandoned. When the meadow portions are extremely small as well as extremely precious, a yearly reallotment of the meadow acreage suggests itself as a matter of equity as well as of common sense.

The introduction of grass seed and clovers marked the end of the dark ages of agriculture. It is the greatest of revolutions, the revolution against the supreme law, the law of the land, the law of diminishing returns and of soil exhaustion.

Go to the ruins of ancient and rich civilizations in Asia Minor, northern Africa, or elsewhere. Look at the unpeopled valleys, at the dead and buried cities, and you can discipher there the promise and the prophecy that the law of soil exhaustion held in store for all of us. It is but the story of an abandoned farm on a gigantic scale. Depleted of humus by constant cropping, land could no longer reward labor and support life; so the people abandoned it. Deserted, it became a desert; the light soil was washed by the rain and blown around by shifting winds.

Only with the introduction of grass seeding did it become possible to keep a sufficient amount of stock not only to maintain the fertility of the soil but steadily to improve it. The soil instead of being taxed year after year under the heavy strain of grain crops, was being renovated by the legumes that gathered nitrogen from the air and stored it on tubercles attached to their roots. The deep roots of the clover penetrated deeper than any plough ever touched. Legumes like alfalfa, producing pound by pound more nutritious fodder than did meadow grass, produced acre by acre two and three times the amount. And when such a field was turned under to make place for a grain crop, the deep and heavy sod, the mass of decaying roots, offered the farmer "virgin" soil, where previously even five bushels of wheat could not be gathered. The tale is a simple one, and it is told.

As for the historical controversy over the village community, the real problem was not touched upon. Deeds and codices were closely examined, but what one could lay his hands upon remained unnoticed. The village exists, the intermixed strips exist, the commons exist. The problem, therefore, was first of all—what is the village community, and secondly, why is it? The question how it developed would then have answered itself. But instead of this search for an explanation we find a quest for the character of the *prototype* of the present village community. This prototype was believed to have been found in an association of freemen that held all land in common, whether arable or not. With the discovery of this prototype, the real problem, *i.e.*, the explanation of the intermixed strips, with resulting compulsory rotation of crops and common pasturage—which constitute a village community—was lost sight of.

Fustel de Coulanges was merciless in his criticism of the documentary evidence submitted by Maurer and others of his school. But just as Maurer failed to see that his very documents prove personal and not common ownership, so did Fustel de Coulanges fail to see anything but private estates in the personally owned strips that were obviously within the boundaries and under the restrictions of a village community. The medieval expressions for the word boundaries are *marcha, marca, finis,* or *terminus* (the Bavarian law says *terminus id est marca*). Fustel de Coulanges quotes: "*dono portionem meam quae est in marca Odradesheim*" and tells us later on: "In the examples given by Maurer, I recognize the existence of the Mark, but a mark which was the same thing as a *villa,* that is a private estate."[1] To my way of thinking one's *portio* within a Marca looks like private ownership within the boundaries of a village community named *Odradesheim.*

If the village community is strange or peculiar among freemen, it is if anything even stranger among serfs, where the landlord could make any arrangement that may have pleased him. Yet this, Fustel de Coulanges entirely overlooks when he sees nothing but a private estate belonging to a monastery in a quota-

[1] Fustel de Coulanges, "The Origin of Property in Land," translated by Margaret Ashley, with an introductory chapter by W. J. Ashley. London, 1891, p. 35.

tion of his which unquestionably indicates a manorial village community. "—*rustici ecclesiae pro quantitate et limitibus contenderent. Ego Hermannus abbas . . . compromissum fuit ut maximus campus per funiculos mensuraretur et cuilibet hubae 12 jugera deputarentur . . . in totidem partes secundus campus et tertius divideretur. . . . Inchoata est ista divisio per Alvinum monachum scribentem et fratzem Bertholdum prepositum et Rudolfum officialem cum funiculis mensurantes.*"[1] How can several *rustici* have strips in the three fields, without having their land intermixed, hence having common pasture on the fallow and stubble and therefore compulsory rotation of crops? Of course, Fustel de Coulanges is right that the document tells us not what Maurer supposed, but indicates rather "a division among tenants carried on by the proprietor."[2] But it tells us quite enough to forget Maurer and pay some attention to the character of the farming indicated by the document. Such quotations, which Maurer misunderstood as common ownership, and Fustel de Coulanges misunderstood as private unrestricted ownership could be found on page after page. So, on the next page: "Another less rich can only give a *huba, but he gives at the same time the portion of the forest to which his huba has a right,*" or in the note on the same page (43): "I will explain elsewhere the meaning of *portio*. All I need say at present is that this word, which occurs more than three hundred times in our authorities, always means a part belonging to an owner. A *portio* is spoken of as *sold,*

[1] *Ibid.*, p. 41.
[2] *Ibid.*, p. 42.

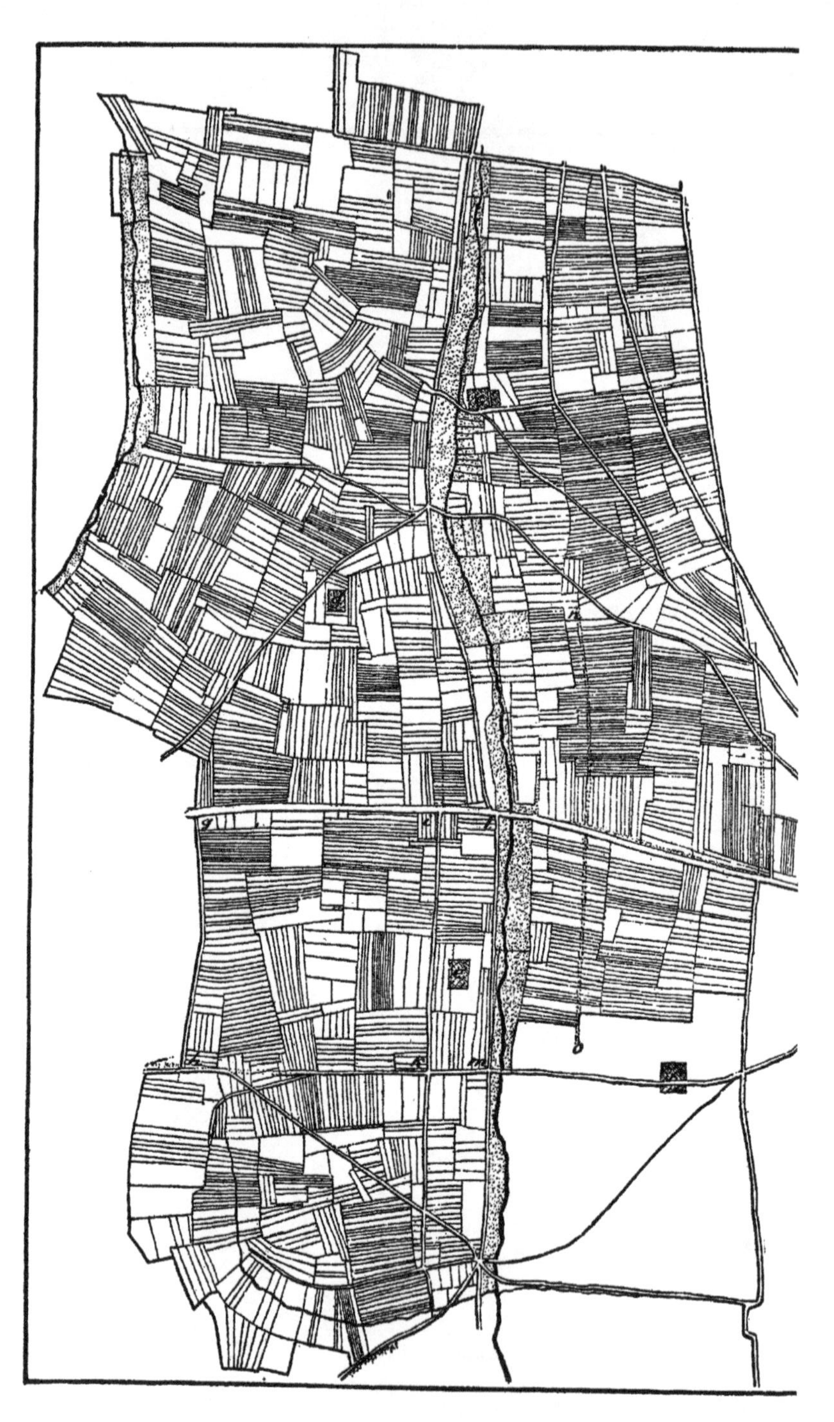

Survey of the village Friedberg in Wetterau, reproduced

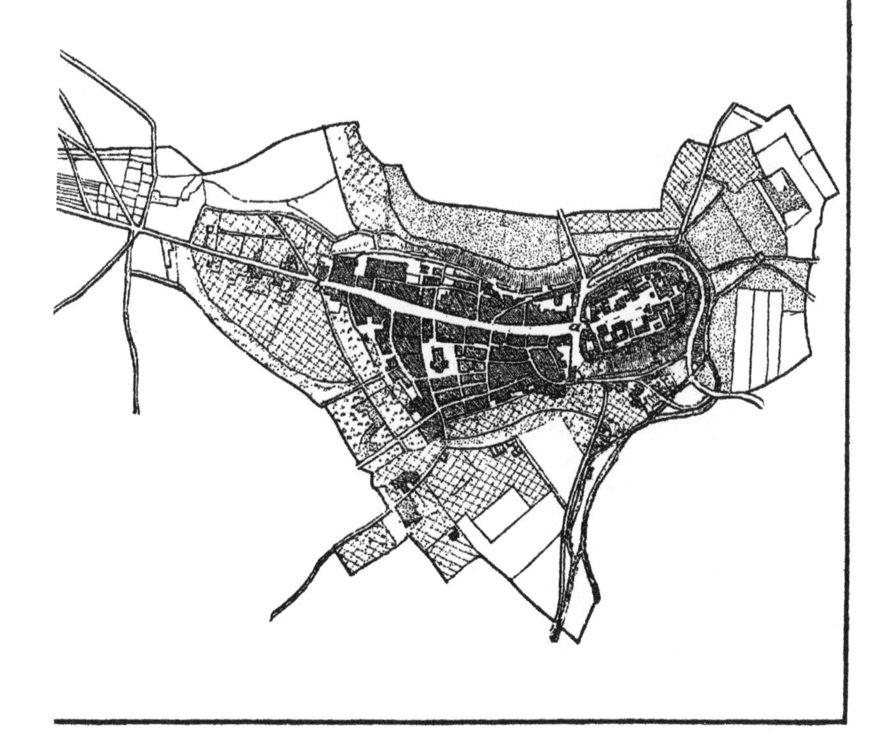

Meitzen's "Siedlung und Agrarwesen," Vol. III, Atlas.

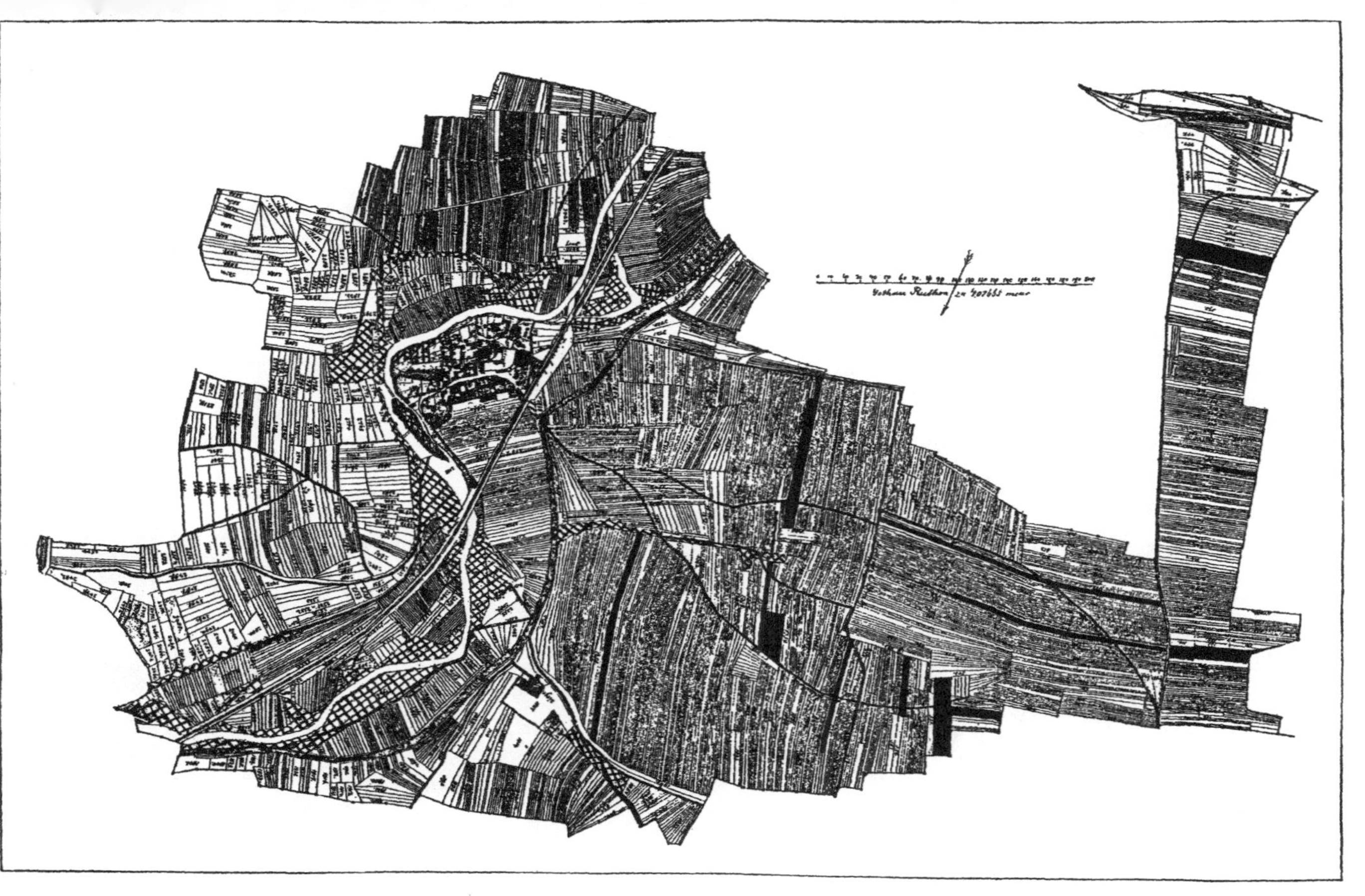

Survey of the village Bishleben, reproduced from August Meitzen's Atlas—Supplement of his work, "Siedelung und Agrarweser der Westgermanen, der Kelten, Römer, Finnen und Slaven." Berlin, 1895.

bequeathed and given." Three hundred land transactions are recorded. The land is sold, bequeathed or given away. The land was personally owned. There is no question about that. But was the use of the land restricted or unrestricted? If unrestricted why are the respective properties in 300 different transactions characterized and legally described not as independent units but as *parts of units?*

I feel very ungrateful in criticizing Fustel de Coulanges. I appreciate the work he has done. And after all I am criticizing him not for what he has done, but for what he failed to do, which is probably unjust.

I must apologize for one other statement. I have emphasized several times that the real problem has been entirely overlooked. This is true but for one exception and that exception is a remark made by W. J. Ashley. In his introductory chapter to Fustel de Coulanges' book, as early as 1891, he wrote:

"In the medieval manor there were two elements, the *seigneurial*—the relations of the tenants to the lord; and the *communal*—the relations of the tenants to one another. The mark theory taught that the seigneurial was grafted on to the communal. The value of the work of M. Fustel de Coulanges and of Mr. Seebohm is in showing that we cannot find a time when the seigneurial element was absent; and also in pointing to reasons, in my opinion conclusive, for connecting that element with the Roman villa. *But the communal element is still an unsolved mystery."* [1]

It is the solution of this mystery that is here submitted to the reader.

[1] Fustel de Coulanges, *op. cit.*, p. xiii.

www.ingramcontent.com/pod-product-compliance
Lightning Source LLC
Chambersburg PA
CBHW021551050725
29163CB00034B/392

* 9 7 8 1 4 1 7 9 0 0 8 7 9 *